Podocytopathy

Contributions to Nephrology

Vol. 183

Series Editor

Claudio Ronco Vicenza

Podocytopathy

Volume Editors

Zhi-Hong Liu Nanjing

John C. He New York, N.Y.

31 figures, 29 in color and 10 tables, 2014

KARGER Basel · Freiburg · Paris · London · New York · Chennai · New Delhi · Bangkok · Beijing · Shanghai · Tokyo · Kuala Lumpur · Singapore · Sydney

Preface

One of the most exciting developments in the field of nephrology in the last decade has been our elucidation of podocyte biology and its role in the pathophysiology of inherited and acquired glomerular diseases. Because of improvements in imaging technologies, the detailed structure of podocytes and their relationship with glomerular function have been directly and dynamically visualized. In addition, advances in genomic technology have contributed to the discovery of several essential podocyte molecules. The function of these key molecules has been well characterized in the context of podocyte biology and glomerular disease. The complex signaling network involved in regulating the podocyte cytoskeleton has also been well studied. Furthermore, recent studies have described a more detailed structure and function of the slit diaphragm, and these findings suggest that crosstalk between podocytes and other glomerular cells is required to maintain a normal glomerular filtration barrier.

A large body of evidence suggests that podocyte injury is a major cause of proteinuric glomerular diseases. Consequently, the term 'podocytopathy' was created to refer to such podocyte diseases. Recent genetic studies suggest that podocyte protein mutations can trigger focal segmental glomerulosclerosis or steroid-resistant focal segmental glomerulosclerosis. In addition, podocyte loss, or podocytopenia, has been shown to be associated with the progression of glomerular diseases. Several mechanisms related to podocyte loss, including apoptosis, detachment, and necroptosis, have also been proposed. Although further studies are required to confirm these mechanisms, podocyte proliferation has been observed in human immunodeficiency virus-associated nephropathy and crescentic glomerulonephritis. Recent evidence suggests that podocytes express phospholipase A_2 receptor, an antigen that is responsible for immune complex deposition in membranous nephropathy. The role of podocytes in other immune-mediated glomerulonephritides, such as IgA nephropathy and lupus nephritis, has also been documented.

Cellular and molecular mechanisms involving podocyte injury have also been recently studied. For instance, mitochondrial dysfunction and endoplasmic reticulum stress have been reported in podocytopathy. Recent evidence also suggests that the generation of excess amounts of reactive oxygen species can result in podocyte apoptosis. In addition, the signaling network leading to podocyte injury has been extensively examined. For example, TGF-β induces podocyte apoptosis in diabetic nephropathy, mTOR, Wnt/β-catenin, and Notch signaling pathways have been shown to mediate podocyte injury in glomerular disease.

Unfortunately, targeted therapies that attenuate or reverse podocyte injury are lacking. Studies suggest that steroids and immunosuppressive medications, such as calcineurin inhibitors, rapamycin and corticosteroids, may exert protective effects that are specifically directed toward podocytes. Vitamin D_3, retinoic acid and triptolide have also been shown to attenuate podocyte injury. Nonetheless, novel therapies that target podocyte injury are urgently needed to treat patients with glomerular disease.

New approaches and tools to study podocyte structure and function have recently been reported. For example, podocyte studies using flies and zebrafish models may provide us with important new information that has not been obtained from mammals. However, podocyte-specific conditional knockout mice and inducible transgenic mice have assisted in generating a large amount of podocyte pathobiology data. The lineage tracing approach may also help us to understand whether podocytes can be regenerated under normal or diseased conditions. Furthermore, a systems biology approach can allow us to generate comprehensive profiles of information at genomic, transcriptomic, proteomic, and metabolomic levels to define podocyte function and failure. The '-omics' studies will certainly play an important role in improving our understanding of podocyte biology and pathology.

For these reasons, it seems critical to have a summary of past contributions in the field of podocyte research, allowing readers to search the relevant references on the pathophysiology of podocytopathies, state-of-art technologies for study of podocytes, and potential new therapeutic approaches for treatment of glomerular disease. In this book, we have assembled several world-renowned experts to summarize the most recent findings and advances in this field, which unfortunately cannot be all-inclusive because of the limited length of this book. We acknowledge the important effort and contributions of the authors who made this summary possible. We also like to acknowledge Dr. Evren Azeloglu for providing us the beautiful cover figure. This publication was supported by the National Basic Research Program of China (Programm 973, Nos. 2012CB517600 and 2012CB517606). Finally, we thank Karger Publishers for their professional editorial assistance, and we hope that readers will enjoy this book.

Zhi-Hong Liu, Nanjing
John C. He, New York, N.Y.

Liu Z-H, He JC (eds): Podocytopathy. Contrib Nephrol. Basel, Karger, 2014, vol 183, pp 1–11
DOI: 10.1159/000360503

Cell Biology of the Podocyte

J. Ashley Jefferson · Stuart J. Shankland

Department of Medicine, University of Washington, Seattle, Wash., USA

Abstract

Background: The terminally differentiated and highly specialized glomerular epithelial cells called podocytes function as a major barrier to protein passaging from the intravascular glomerular capillaries to the extravascular urinary space. However, once injured in disease, increased passage of proteins leads to proteinuria, and a decrease in podocyte number underlies progressive glomerular scarring. ***Summary:*** Numerous proteins specific to podocytes enable their normal functions, including those that comprise the slit diaphragm to act as a size, charge and shape barrier, a rich actin cytoskeleton that enables mobility, and the production and secretion of growth factors required for normal glomerular endothelial cell health. When injured in podocyte diseases such as focal segmental glomerulosclerosis, minimal change disease, membranous nephropathy, diabetic kidney disease and others, several of these normal functions are disrupted, leading to changes in histological appearance, structure and function. These are typically manifest clinically by proteinuria and a decline in kidney function. ***Key Messages:*** Because of podocyte's inability to adequately proliferate, a decline in their number follows when cells undergo apoptosis, detachment, necrosis and altered autophagy in response to injury. This leads to progressive glomerular scarring. These mechanisms will be discussed in this chapter. Alterations in key slit diaphragm proteins lead to proteinuria, which will also be discussed.

Podocytes are terminally differentiated epithelial cells lining the outer aspect of the glomerular capillaries. Their complex ultrastructure consists of a cell body, from which extend long branching cellular processes, comprising primary and then secondary processes. These end in foot processes, which attach podocytes to the underlying glomerular basement membrane (GBM). Podocytes serve several critical biological functions. First, highly specialized cell junctions between adjacent foot processes are bridged by slit diaphragms which allow ultrafiltrate to pass but limit the passage of critical proteins such as albumin [1]. They act as a size, shape and charge barrier to the passage of protein from the intravascular compartment of the underlying capillary network to the extravascular urinary space. Second, the rich actin cytoskeleton enables podocytes to actively maintain the integrity of the underlying glomerular capillaries [2]. Third, podocytes make extracellular matrix proteins such as laminin β_2 and the collagen $\alpha_3,\alpha_4,\alpha_5$ (IV) network required for the development of the normal GBM, and likely thereafter, for normal GBM maintenance. Finally, podocytes secrete critical survival factors for neighboring glomerular endothelial cells, such as VEGF and angiopoietin-1 [3]. Following injury in primary glomerular diseases such as focal segmental glomerulosclerosis (FSGS), membranous nephropathy and minimal change disease [4], as well as secondary glomerular diseases such as diabetic nephropathy, one or more of these biological functions are reduced or absent, leading to the characteristic histological and clinical manifestations described below.

Responses of Podocytes to Injury

Although each glomerular disease has unique mechanisms by which they induce podocyte injury, there are several common responses to injury that will be discussed below (fig. 1).

Podocyte Injury and Proteinuria

The clinical signature of podocyte injury is proteinuria, predominantly albuminuria. As discussed elsewhere in this series, although the magnitude of proteinuria varies, podocyte diseases must be considered when patients present with nephrotic range proteinuria (≥3.5 g/24 h). Several mechanisms underlie proteinuria following podocyte injury:

(i) *Changes in slit diaphragm proteins*: Several 'podocyte-specific' proteins are concentrated in the slit diaphragm, whose primary function is to form and support the sieve-like structure of this modified tight junction [1]. These filtration slits are 40 nm wide between adjacent processes, slightly smaller than the size of albumin. Nephrin is the protein that forms the major backbone, but requires complexing with several other proteins such as podocin and NEPH for normal function, as well as several adapter proteins such as CD2AP [5]. The slit diaphragm structure is size-selective, although a negative charge might also limit albumin passage. Abnormalities of slit diaphragm proteins can arise from several events in disease. First, a decrease in absolute levels of one or more of these proteins in acquired glomerular diseases disrupts the overall barrier function. Second, mutations in genes encoding nephrin, podocin, TRPC6 or CD2AP oc-

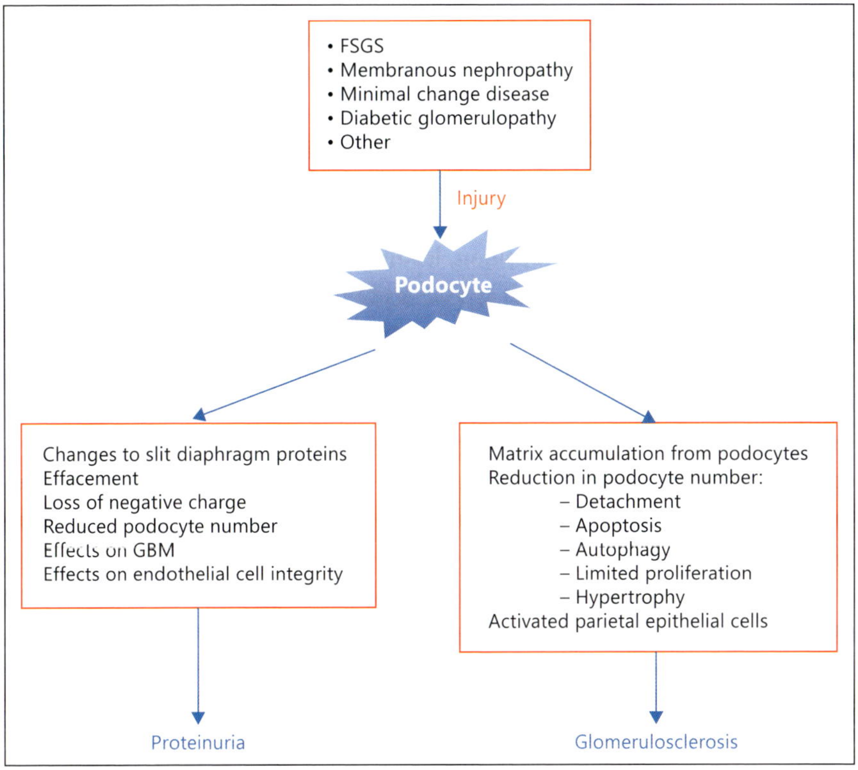

Fig. 1. Response to podocyte injury. Several diseases target podocytes, and induce injury. The events leading to proteinuria and glomerulosclerosis are summarized.

cur in several forms of congenital and hereditary glomerular diseases [6]. Third, disease-induced changes in the subcellular location of nephrin or podocin prevent their normal function [7].

(ii) *Effacement*: The characteristic ultrastructural finding of podocyte injury is called effacement, which is seen on electron microscopy as a flattening of foot processes. This is an active process mediated by changes in the actin cytoskeleton. Effacement leads to distortions and loss of the slit diaphragm impairing its normal barrier and signaling function. See below for more detailed discussion.

(iii) *Loss of podocyte negative charge*: Podocytes are negatively charged due to anionic proteins such as podocalyxin. Loss of this negative charge, due to alterations in these proteins, limits the charge selective barrier capacity of podocytes.

(iv) *Reduced podocyte number*: Although reduced podocyte number is typically considered as a major cause of glomerulosclerosis, their absence leads to uncovered areas on the outer aspect of the GBM, through which albumin and other proteins might easily traffic.

when key autophagic proteins such as atg5 are deleted in podocytes, damaged mitochondria and oxidized products accumulate, leading to worsening glomerular disease [27].

(iv) Other forms of death: Several other models of podocyte death have been described. Cell lysis due to necrosis occurs in a subset of podocytes in membranous nephropathy following antibody-induced complement activation and cell activation. Anoikis, necroptosis and mitotic catastrophe have all been documented in podocytes.

(v) Limited ability to proliferate: The aforementioned mechanisms lead to a loss of podocytes, and consequently a decrease in their number. Podocytes are post-mitotic terminally differentiated epithelial cells, and in contrast to other kidney cells, adult podocytes have a very low proliferative capacity. In order for a cell to proliferate, it has to enter the cell cycle. In the G1 phase this requires cyclin-dependent kinase 4 (Cdk4) to be activated by D-type cyclins. In podocytes, cyclin D1 increases following injury, and binds to Cdk4 [28]. However, the Cdk inhibitors p21 and p27 are also increased, and by complexing with cyclin D-Cdk4, limit their activity [29]. Both inhibitors also bind to and inhibit cyclin A-Cdk2, required for DNA synthesis [29]. Together, the increase in these Cdk inhibitors is a major reason why podocytes cannot undergo adequate DNA synthesis. Moreover, despite podocytes ability to increase cyclins B1 and B2 required for mitosis, albeit at very low levels, podocytes reaching mitosis seem to undergo 'mitotic catastrophe', thereby failing to complete cytokinesis [30]. Thus, the predominantly inhibitory influences on the cell cycle by Cdk inhibitors limit/inhibit podocyte proliferation, and in certain diseases leads to loss greater than regeneration. The overall result is reduced podocyte number.

Although the overall paradigm related to podocyte proliferation is that they typically are unable to proliferate in adults, several groups have suggested that they might proliferate under two conditions: crescentic glomerulonephritis and collapsing glomerulopathy. The data for the former derives largely from experimental studies in podocyte reporter mice, showing that labeled cells (i.e. podocytes) are detected in the crescents [31]. While this might be factual, subsequent studies have shown that the vast majority of cells in glomerular crescents are proliferating parietal epithelial cells [32]. The data proving that podocytes are the major cell proliferating in pseudo-crescents of collapsing glomerulopathy is indirect. Studies showed that the cells in pseudo-crescents from human biopsies no longer express several well-established podocyte markers such as WT-1 [33, 34]. This led to the notion that the terminally differentiated podocytes had undergone dedifferentiation in order to engage the cell cycle [33]. This paradigm has never been proven functionally. Rather, a large body of experimental and clinical evidence now suggests that parietal epithelial cells are the major cell type in pseudo-crescents, and that a subset of these are stem/progenitor cells for podocytes [35]. Finally, recent studies have shed light on cells of renin lineage as potential stem/progenitor cells for podocytes in experimental FSGS [36].

Taken together, adult podocytes have a very limited ability to proliferate, if at all, whereas the parietal epithelial cell readily undergoes proliferation, and likely accounts for crescents and pseudo-crescents in certain forms of glomerular disease. Indeed, in progressive glomerular diseases, podocyte loss cannot be matched by podocyte proliferation, which leads to overall podocyte depletion, with consequent glomerulosclerosis.

(vi) Hypertrophy: Due to the inability of podocytes to adequately proliferate, several studies have focused on podocyte hypertrophy as a mechanism to 'accommodate' to the increased glomerular volume accompanying several glomerular disorders including aging. Hypertrophy is defined as an increase in cell size due to an increased protein-to-DNA ratio. One might imagine that following a loss of podocytes, their neighbors increase in size in order to 'cover' the now bare GBM. A seminal study by Wiggins et al. [37] showed that in aging rats, podocytes hypertrophied up to 2.2 times their normal size. Wiggins et al. noted 5 stages of podocyte hypertrophy, where stage 1 was normal, stage 2 was accompanied by non-stressed hypertrophy, without proteinuria, stage 3 was described as adaptive hypertrophy, where podocyte function remained normal, stage 4 was characterized by decompensated podocyte hypertrophy, where normal podocyte proteins such as nephrin were reduced, foot process width increased, and proteinuria was increased, and stage 5 was accompanied by podocyte depletion, where the failure of podocytes to meet demand lead to increased podocyte loss, and consequent glomerulosclerosis. The mechanisms of podocyte hypertrophy are likely multifactorial. Studies in mouse podocyte exposed to mechanical stretch showed that the resultant hypertrophy is dependent on the expression of specific cell cycle proteins and signaling pathways [38]. The results showed that stretch did not induce podocyte hypertrophy in the absence of the Cdk inhibitor p21 [38]. The Cdk inhibitor p27 had no impact on podocyte hypertrophy. Finally, the signaling pathways ERK1/2 and AKT are necessary for stretch-induced podocyte hypertrophy [38].

Activation of Parietal Epithelial Cells (PECs)

There is increasing data showing that although the primary injury in diseases such as FSGS is to podocytes, the neighboring PECs become secondarily activated. Although the mechanisms are not understood, a subset of PECs begin to express CD44 [39]. Two events contribute to glomerulosclerosis. First, CD44-expressing PECs begin to produce increased amounts of extracellular matrix considered unique to Bowman's basement membrane. This might lead to the

The Podocyte's Limited Proliferative Capacity: Role of Cell Cycle Proteins

Experimental evidence from the passive Heymann nephritis (PHN) model of membranous nephropathy, induced by the administration of an antibody reactive against the Fx1A antigen on the rat podocyte, has raised the paradigm that defective DNA synthesis or mitosis and cytokinesis may underlie the well-appreciated lack of podocyte proliferation. In the acute phase of disease, mitotic figures and an increase in polynucleated cells have been described together with increased protein levels for cyclin A and Cdk2, consistent with cell cycle activation [7]. However, there is only a very limited increase in DNA synthesis, yet podocyte number decreases over time. Together, this suggests the presence of an inhibitor(s) to cell cycle progression as demonstrated by increased protein expression of the Cdk inhibitors $p21^{Cip1}$ and $p27^{Kip1}$ [7]. Administering the mitogen bFGF to PHN rats further attenuates the increase in $p21^{Cip1}$ resulting in increased podocyte DNA synthesis and ploidy. In addition, Petermann et al. [11] reported an upregulation of the M-phase cell cycle proteins Cdc2 and cyclin B in PHN podocytes. The authors speculated on a disturbed cytokinesis to be ultimately responsible for the development of ploidy and the observed lack of podocyte proliferation.

In line with these findings are cell culture experiments from Pippin et al. [12] that further explored the inability of podocytes to proliferate following C5b-9-induced injury. When podocytes are exposed to a sublytic C5b-9 attack in vitro, they engage the cell cycle. However, similar to their in vivo counterparts, cells do not enter mitosis, suggesting a cell cycle block in the G_2/M transition checkpoint mediated by increased checkpoint kinase-1 and -2 protein levels. This response is typical following DNA damage, thus arresting cells at the G_2/M boundary of the cell cycle. The mechanism by which DNA damage occurs in podocytes is still not well understood, but may comprise the generation of reactive oxygen radicals.

The use of $p21^{Cip1}$ knockout mice established the decisive role of $p21^{Cip1}$ in limiting the proliferative response of podocytes [13, 14]. While the absence of $p21^{Cip1}$ does not affect glomerulogenesis or glomerular function, the absence of $p21^{Cip1}$ promotes podocyte dedifferentiation and proliferation in response to injury in the nephrotoxic nephritis model. Marshall et al. [15] reported a similar response in the adriamycin model of focal segmental glomerulosclerosis (FSGS) in mice. Additional in vitro data link the pro-apoptotic effect of TGF-β_1 in podocytes to $p21^{Cip1}$ [16].

The inability of podocytes to proliferate in response to injury and subsequent podocyte loss has been confirmed in human diseases where similar underlying mechanisms have been observed. In humans, normal quiescent podocytes express $p27^{Kip1}$ and $p57^{Kip2}$, and immunostaining for these proteins is maintained in the podocyte diseases such as minimal change diseases and membranous glomerulopathy without any evidence of podocyte proliferation [17]. More importantly, expression of both the Cdk inhibitor proteins $p27^{Kip1}$ and $p57^{Kip2}$ is uniformly decreased and accompanied by the de novo protein expression of $p21^{Cip1}$ in diseases characterized by podocyte proliferation including cellular FSGS, collapsing glomerulopathy and HIV-associated nephropathy (HIVAN).

The Podocyte's Limited Proliferative Capacity: Role of Mechanical Stretch

The unique anatomic localization of podocytes at the outer surface of the glomerular capillaries bathed in primary urine where they are attached only to the underlying GBM and to its podocyte neighbor via the slit diaphragm renders these cells at high risk for detachment during the process of cellular division. In addition, podocytes are also exposed to mechanical stress as seen in glomerular hypertension. Lowering intraglomerular pressure reduces disease progression in several glomerular diseases, including diabetic nephropathy [18]. Glomerular hypertension not only leads to glomerular hypertrophy, but the resultant mechanical stretch also injures all three resident glomerular cell types. Whereas applying mechanical stretch to cultured mesangial cells increases proliferation [19], the opposite response is observed in podocytes due to decreased levels of cyclins D1, A, and B1 and Cdc2 [20]. Simultaneously, stretch causes a rapid increase in $p21^{Cip1}$, followed by increases in $p27^{Kip1}$ at 24 h and $p57^{Kip2}$ at 72 h. In contrast to the growth arrest in wild-type cultured podocytes exposed to stretch, $p21^{Cip1-/-}$ podocytes exposed to stretch continued to proliferate, suggesting a role for $p21^{Cip1}$ in the inability of the podocyte to progress through the cell cycle in response to stretch.

Podocyte Proliferation in Response to Injury: Where It Occurs

As outlined above, the majority of experimental models of podocyte injury and human podocyte diseases are not associated with proliferation. However, under certain conditions, podocyte proliferation has been reported including collapsing glomerulopathy, cellular FSGS and HIVAN [21, 22] based on increased expression of cyclin A, the proliferation marker Ki-67, and reduced expression of $p27^{Kip1}$ and $p57^{Kip2}$ [17, 22]. However, more recent studies have implicated parietal epithelial cells to be the glomerular cell type that proliferates in these

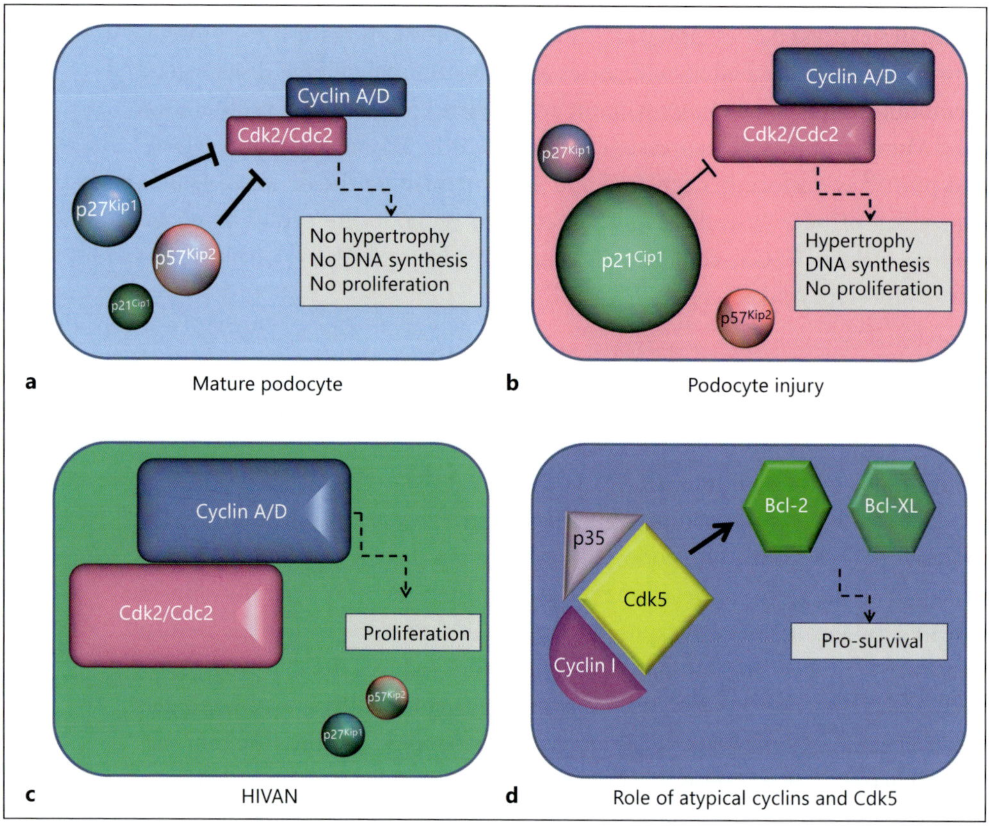

Fig. 1. Cell cycle control in the podocyte. In mature podocytes, high expression levels of $p27^{Kip1}$ and $p57^{Kip2}$ inhibit cyclins and Cdk (e.g. Cdk2, Cdc2). Thus, cell cycle progression is stalled and protein and DNA synthesis is hampered to circumvent cellular hypertrophy, polyploidy and proliferation (**a**). After injury the expression levels of $p27^{Kip1}$ and $p57^{Kip2}$ decline, whereas the expression of $p21^{Cip1}$ increases to maintain G1/S arrest. However, in the presence of substantial damage, $p21^{Cip1}$ cannot sustain the G1/S checkpoint, resulting in hypertrophy and polyploidy. Proliferation may still be precluded by a G2/M arrest mediated by checkpoint kinases (**b**). In advanced podocyte disease and HIVAN the expression of all cell cycle inhibitors is decreased, resulting in unopposed cyclin and Cdk activity. The podocytes proliferate (**c**, **d**). Neurons and podocytes host the atypical Cdk5 and its activators cyclin I and p35. Activation of Cdk5 after injury leads to increased expression and stabilization of the pro-survival proteins Bcl-2 and Bcl-XL.

diseases. Interestingly, progression to end-stage renal failure is found to be increased in glomerular diseases characterized by podocyte proliferation [23, 24].

The pathogenesis of HIV-associated nephropathy has been further delineated using transgenic mice expressing HIV-1 genes [25–27]. HIV-1 induces loss of contact inhibition in podocytes [28] and expression of cyclin D1 and phosphorylation of pRb [29]. The cells become dedifferentiated and lose expression of podocyte-specific proteins [25, 30]. Treatment with highly effective antiretroviral regimens not only suppresses HIV load but is also capable of reversing the glomerular changes observed in HIV nephropathy [31], suggesting that once the HI virus is cleared, the podocyte is capable of exiting the cell cycle again to re-differentiate to its mature phenotype.

Cell Cycle Proteins: Podocyte Hypertrophy

Hypertrophy is defined as an increase in cell size due to a high ratio of cell protein to the DNA content. Several lines of evidence show that in the absence of adequate proliferation, podocytes undergo hypertrophy in an attempt to cover the underlying GBM rendered denuded by disease-induced loss. Wiggins and colleagues [1] showed different stages of podocyte hypertrophy. Initially this was adaptive and compensatory. Later hypertrophy was maladaptive. In addition to growth factors and other regulators, changes to cell cycle proteins might underlie podocyte hypertrophy. When cultured mouse podocytes were exposed to mechanical stretch they hypertrophied, accompanied by higher levels for the Cdk inhibitors p21, p27 and p57 [11]. This was accompanied by a decrease in the activity for Cdk2. Follow-up studies using cells devoid of Cdk inhibitors showed that p21 was essential for the arrest of the cell cycle machinery (i.e. prevented DNA synthesis), while the cell increased its protein content in the presence of mitogens [20].

Cell Cycle Proteins: Effects beyond Proliferation

The traditional view of cell cycle proteins in podocytes and other cells has been challenged by the discovery of 'atypical' cyclin proteins and dependent kinases. In addition to its well-defined regulatory function on cellular proliferation and subsequently differentiation, certain cell cycle proteins are also critically involved in setting the threshold for podocyte survival/apoptosis. Among the atypical cyclins is cyclin I and its associated kinase Cdk5 [32]. Cyclin I and Cdk5 are predominantly expressed in terminally differentiated cells including neurons and podocytes. The absence of cyclin I in mice does not affect glomerulogenesis or function, but renders podocytes more susceptible to apoptosis following injury [33]. A potential mechanism includes a decreased expression and protein stability of $p21^{Cip1}$. Very little is known about the function of Cdk5 in the kidney. In addition to cyclin I, Cdk5 is also activated by the non-cyclin

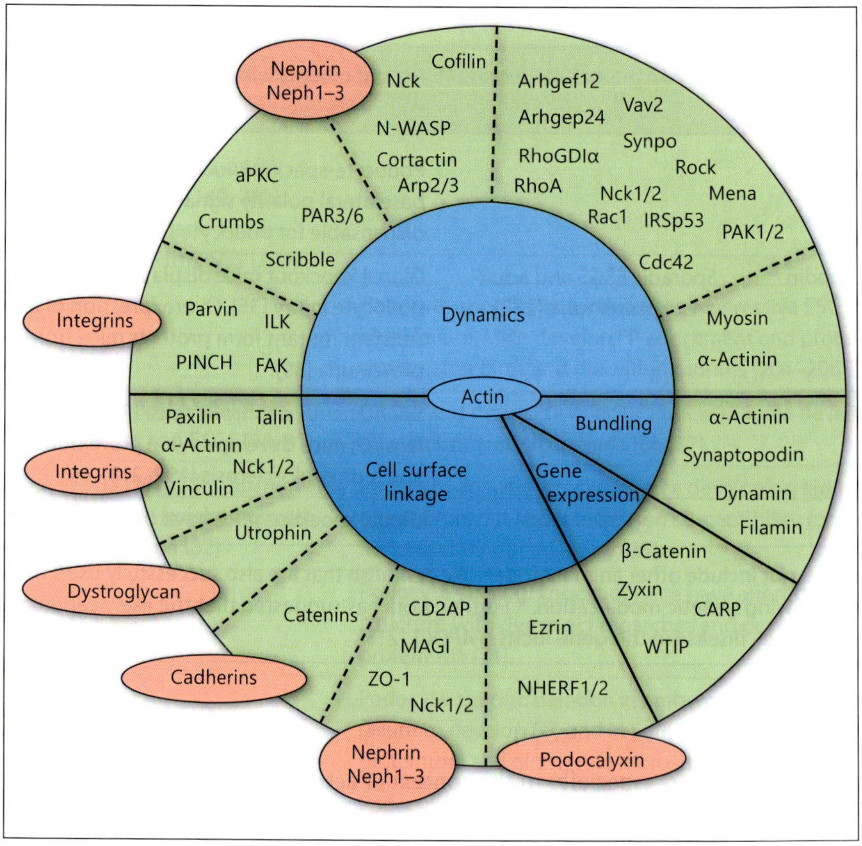

Fig. 1. The multiple functions of the actin cytoskeleton and the big variety of upstream regulators present in podocytes. Major actin functions (blue circle) are the regulation of cell dynamics including cell motility and contraction. Because actin is linked to cell surface complexes like the SD and focal contacts it affects signal transduction that originates from these structures. Polymeric actin is bundled as filamentous forms with different lengths and degree of branching. Actin also communicates with the nucleus to regulate gene expression. Cytoplasmic proteins directly or indirectly bind actin (green ring) and regulate the structure and dynamics of the actin cytoskeleton, some in communication and contact with cell surface proteins and complexes (red oval in periphery).

congenital nephropathy. Homozygous mice die from kidney failure within 2 weeks after birth [42, 43]. In contrast, podocyte-specific deletion of Rac1 or RhoA does not cause a glomerular phenotype [42, 43]. These mouse models suggest that Cdc42, but not Rac1 or RhoA, is critical for establishing the podocyte cytoarchitecture during development and for the maintenance of the glomerular filtration barrier under physiologic conditions. Interestingly, Rac1–/– mice are protected from PS-induced acute podocyte injury, while long-term injury via chronic hypertension exacerbates albuminuria and glomerular injury in Rac1–/–

mice [43]. Therefore, Rac1 seems to be dispensable in the physiologic state, but it has beneficial or deleterious effects depending on the context of podocyte impairment. Cdc42 might have similar 'janus face'-like actions since blockade of Mena, which is downstream of Cdc42 and required for Cdc42-induced filopodia formation, with FP4-Mito in the lipopolysaccharide (LPS) model blocks proteinuria [38]. Furthermore, Rac1 and Cdc42 target p21-activated kinase (PAK). PAK family of serine/threonine kinases are key regulators of actin dynamics, but its precise mechanism of action is still unclear. It has been shown that PAK1 activation in cultured podocytes induces actin rearrangement [44] and that PAK1 inhibition by IPA-3 protects podocytes from glucose-induced injury in vitro [45].

Podocyte-specific gain-of-function studies in mice added further light as well as confusion to the story. Overexpression of constitutive active Rac1 causes rapid-onset proteinuria and FP effacement within only 2 days of transgene induction [46]. However, the proteinuria is not durable and gradually decreases within 1 month even in the presence of the transgene. This model indicates that overactivation of Rac1 in adult podocytes induces rapid changes in podocyte morphology, but co-activation of other, so far unidentified mechanisms eventually compensate for the injury. Similarly, the podocyte-specific overexpression of constitutive active RhoA also causes proteinuria and podocyte injury in adult mice [47, 48]. Low level transgene expression resulted in an MCD-like phenotype, while high expression levels caused FSGS-like alterations, and the phenotypes were partially reversible upon removal of the transgene inducer [47]. Proteinuria developed only 2 weeks after transgene induction, whereas in vitro effects caused by the same mutant included rapid cytoskeletal changes within hours. Surprisingly, also the overexpression of a dominant-negative RhoA mutant induces albuminuria and FP effacement [48]. The mechanisms of action between constitutive active versus dominant-negative mutant forms seem to differ, since the active form promotes podocyte apoptosis, while the dominant negative form does not affect cell survival [48]. A number of treatment studies suggest that overactivation of RhoA might be detrimental for podocytes. A major downstream effector of RhoA is Rho kinase (ROCK), and interestingly inhibition of ROCK by Y27632, fausadil or SAR407899 inhibits renal injury and proteinuria in various animal models for experimental nephropathies [49–53]. Furthermore, ROCK inhibitors block actin rearrangement induced by mechanistic stress [54–56]. These studies indicate that RhoA appears to play an important role in maintaining the integrity of the glomerular filter under basal conditions, but enhancement of RhoA activity above basal levels promotes podocyte injury.

Studies on regulators of RhoGTPase activities or their status of GDP/GTP loading support the important role of RhoGTPases in podocytes. Arhgap24, a

RhoA-activated Rac1-GTPase activating protein (GAP), confers balance between RhoA and Rac1 signaling, and increased Rac1 and Cdc42 activities can be measured in Arhgap24 knockdown podocytes [57]. Mutations resulting in reduced enzymatic activity of Arhgap24 have been identified in patients with FSGS [57], indicating that Rac1/Cdc42 overactivation in the absence of Arhgap24 might result in increased podocyte motility and injury. Furthermore, human genetic studies showed that *ARHGDIA*, which encodes for RhoGDIα, is mutated in patients with FSGS. These mutations cause a loss of RhoA binding, increase active Rac1 and Cdc42, and enhance podocyte migration in vitro, which can be reversed by Rac1 inhibitors. These findings support the earlier in vivo studies showing that RhoGDIα knockout mice develop proteinuria and progressive kidney injury [29].

Overall these studies indicate that activation of RhoGTPases in podocytes must be tightly regulated, and that inappropriate activation of RhoA is pathopysiologically important. Even smallest changes in activity and expression levels might result in different phenotypes so that activation or inhibition of a particular RhoGTPase should not be labeled as good or bad. Since these enzymes regulate each other, their fine balance might be important and change during the course of podocyte injury. This has been described in the rat model of puromycin aminonucleoside (PAN) nephrosis, where RhoA activity is elevated at day 7 (the peak of proteinuria), while Rac1 activity increases only after 14 days, when the recovery process has started [58]. Also, complement-induced podocyte injury has been shown to affect the RhoGTPases differently and to alter the balance of their activities [59].

Podocyte Injury: Too Much or Too Little Motility?

It has been postulated that RhoA-mediated formation and stabilization of actin stress fibers in podocytes causes a stationary phenotype, while Rac1 and Cdc42 induce lamellipodia and filopodia, respectively, and thereby increase cell motility [19]. Due to the presence of stress fibers in healthy podocytes it has been suggested that RhoA activity confers a healthy stationary phenotype, whereas a predominant activation of Rac1/Cdc42 mediates a disease-associated motile phenotype resulting in instable or retracted FP as found in effaced podocytes. Functional studies of upstream regulators of RhoGTPases like synaptopodin have clearly supported such a hypothesis. The loss of synaptopodin in cultured cells decreases RhoA levels causing a loss of actin stress fibers, and elevates Rac1/Cdc42 activity resulting in increased lamellipodia and filopodia formation and cell motility [35, 37, 38, 60]. Since a reduction in synaptopodin protein levels in mice has been shown to cause proteinuria [39], it has been postulated that this effect is due to the described changes in cell motility. Other in vitro studies have supported this model, where established stimuli that lead to the development of FP effacement in vivo (like LPS and PAN) cause hypermotility [61, 62].

Furthermore, several studies have indicated that Rac1-mediated hypermotility induces podocyte injury and proteinuria. RhoGDIα-null mice have elevated Rac1 activity and develop proteinuria [29], and patients with mutations in *ARHGDIA* have increased Rac1 and Cdc42 activity, and develop podocyte FP effacement and proteinuria [63, 64]. Furthermore, Rac1 can activate PAK1 in podocytes and thereby the mineralocorticoid receptors which accumulate in the nucleus [65]. In this in vivo model the delivery of the Rac1 inhibitor NSC23766 reduces mineralocorticoid receptor activity in podocytes, glomerular injury and proteinuria [65]. Furthermore, the activation of Rac1 signaling in podocytes contributes to proteinuria in HIV-associated nephropathy (HIVAN) [66]. Mutations in *Arhgap24* result in increased Rac1 activation in vitro and correlate with podocyte injury in FSGS patients [57]. These studies suggest that inappropriate Rac1 activation plays a pathogenic role in podocytes. A recent treatment study in 5/6-nephrectomized mice with the Rac1 inhibitor EHT1846 further indicates that Rac1 is a mediator of podocyte dysfunction in kidney disease [53].

While pathological implications of Rac1/Cdc42 activation in podocytes have become clear over the past years, in vivo evidence that hypomotility is beneficial for podocytes and glomerular filter function is still missing. Expression of dominant active RhoA in cultured podocytes increases the number of stress fibers and focal adhesions accompanied by reduced cell motility, while the expression of dominant negative RhoA causes the loss of actin stress fibers and a focal adhesions and increased cell motility [48]. However, as described above, the expression of both RhoA mutant forms results in the same glomerular pathology. RhoA activation in cultured podocytes induces cell contraction and loss of podocyte processes [58], and ROCK inhibitors reduce renal injury and proteinuria in various animal models [49, 50]. These findings indicate that increased stress fiber content and reduced cell motility can also have harmful effects and cause podocyte injury, as does decreased stress fiber content and increased motility. The scenario that podocytes with stabilized actin fibers are more vulnerable to injury has been proposed for the mechanism underlying hereditary forms of FSGS caused by mutations in α-actinin 4 [67], an actin-bundling protein associated with stress fibers in podocytes [35]. Missense mutations result in a higher affinity of the mutant protein for stress fibers and a stiffer actin cytoskeleton [68, 69].

RhoA appears to play an important role in maintaining the integrity of the filter barrier under physiologic conditions, but enhancement of RhoA activity above basal levels promotes podocyte injury. Overall, as in most biological systems like protein kinase activation in signal transduction pathways, molecules do not always function as switch on/off buttons. It is important how many molecules over what time in what precise location within the cell are activated [70]. It is clear that for proper podocyte plasticity a tight and coordinate regulation of RhoGTPases is required, and that the podocyte motility hypothesis might be oversimplified. Contradictory experimental findings could be explained by differences in target structures, as RhoGTPases have not only been shown to regulate the actin cytoskeleton, but also other cellular processes like cell survival [30]. It is worth noting that the observation of actin rearrangement in podocyte cultures might not always be accompanied by changes in cell migration. For example, N-WASP knockout mice develop proteinuria and altered FP structure and isolated podocytes from these mice show significantly impaired actin reorganization dynamics, including dorsal ruffles. However, the basic actin-based process of cell migration is not affected in these cells [71]. Furthermore, one should be especially cautious when interpreting phenotypic findings in transgenic mouse models with podocyte-specific overexpression of proteins, like the described models for the RhoGTPase mutant forms. Since Rac1 can bind many different proteins, including CD2AP, it is possible that phenotypic changes in mice with podocyte-specific overexpression of Rac1 are not due to Rac1's positive effects on lamellipodia formation and cell migration, but for example caused by rearranging the localization of CD2AP [72], which itself is a key regulator of signal transduction at the SD and actin organization in FP [73–75]. Therefore, these transgenic mouse models do not necessarily provide direct proof that the increased Rac1-mediated podocyte migration, as observed in vitro, is the underlying cause for changes in FP morphology and the induction of proteinuria in vivo. Finally, it is likely that the overexpression of any protein per se acts as a podocyte stressor that induces a compensatory response including the 'classic' steps of podocyte injury, i.e. a rearrangement of the actin cytoskeleton resulting in alterations of the sophisticated FP structure and cell morphology and changes in the selectivity of the glomerular filter. Therefore, the detection of proteinuria in such mouse models might not always be directly caused by the gain or loss of function of this particular protein or signaling mechanism, but by an attempt of the podocyte to protect itself from this artificial stress situation.

Stress Fibers and Podocyte Contraction

Actin cannot only regulate cell dynamics by changing the length of its fibers and the degree of its interconnections, but also by serving as rails for motor proteins thereby mediating cell contraction. Podocytes have stress fibers that are actin-myosin-based contractile machineries and are associated with focal adhesions [5]. Stress fibers mediate sustained and fine-tuned contraction within cells as well as cell motility [76]. Although stress fibers have been described in many cultured cells, like fibroblastic and endothelial cells, only a few cell types have actually been shown to have stress fibers in vivo; endothelial cells of blood vessels and podocytes are among these exceptions [5, 77]. The molecular composition of stress fibers is similar to that of sarcomers, the contractile units of striated myocytes, where actin filaments are bundled by α-actinin with intercalated myosin II motor filaments, and sets of regulatory proteins including tropomyosin and filamin [76]. Podocytes express non-muscle α-actinin 1 and 4 and non-muscle myosin IIa (table 1), and in FP synaptopodin takes over the function of tropomyosin [78], which is only found in the cell body of podocytes [5]. Phosphorylation of myosin light chain is the primary event in the regulation of stress fiber contractility [79], and ROCK as well as intracellular calcium levels are the main upstream regulators of myosin light chain activity, thereby incorporating RhoGTPase and calcium signaling with actin dynamics [80]. Surprisingly, very little is known about the molecular pathways that regulate actin fiber contraction in podocytes. Mechanistic studies are rare, which might be due to the complexity of measuring cell contraction and cellular forces in highly differentiated cells like podocytes in vitro (when compared to 'simple' brick-shaped cells like cardiomyocytes). It is likely that as in myocytes, calcium does not only have second messenger function towards signal mediators like the calcineurin/synaptopodin/NFAT axis [39, 81], but also regulates myosin II-dependent contractility of podocyte FP. Furthermore, many ABPs like α-actinin are calcium-binding proteins and their ability to interact with actin is regulated by changes in intracellular calcium levels.

Human genetic studies have unveiled the importance of the actin-myosin machinery for proper podocyte function. Several mutations in *MYH9* (which encodes for non-muscle myosin heavy chain IIa or Myo2a) have been causatively linked to the development of Epstein and Fechtner syndromes, which are both characterized by nephritis [82]. Furthermore, mice with podocyte-specific *MYH9* deletion develop normally but are predisposed to adriamycin-induced podocyte injury [83]. Also, mutations in *MYO1E*, which encodes for myosin 1e, have been associated with familial FSGS [84, 85]. Global as well as podocyte-specific Myo1e deletions in mice result in FP effacement and proteinuria [86, 87]. Depletion of Myo1c in zebrafish causes abnormal glomerular filter function and ultrastructure and a loss of the podocyte FP [88]. Mechanistically, Myo1c regulates the transport of nephrin and Neph1 to the cell surface [89]. However, since class I myosins like Myo1c and Myo1e do not form filaments but in their monomeric form bind molecular cargoes, it is likely that these mutations reflected changes in other important features of the actin cytoskeleton like vesicle transport and/or endocytosis, as recently indicated for Myo1e in cultured podocytes [90].

Overall, the contractile contribution of the podocyte to the glomerular capillary is controversial. However, recent in vitro studies using novel techniques of measuring cell contraction and forces indicate potential implications for stress fiber's in vivo ability to regulate glomerular dynamic and permeability characteristics [91]. Podocytes are mechanosensitive and their ability to respond to shear stress and to contract has been described [54]. Furthermore, angiotensin II elevates intracellular calcium levels and induces contraction in podocytes [92]. Combined, these few studies indicate that contraction-mediated adaptations to external stimuli are physiologically relevant. In the presence of mutations that lead to a relaxation of cytoskeletal tension, podocytes inadequately counteract physiological forces. Changes in contractility most likely involve signaling pathways that have been described in other contractile cells like smooth muscle cells, but to date they are poorly understood in podocytes. Of note, podocytes have been shown to express several proteins like smoothelin and calponin [91] (table 1), which are known to regulate contraction of smooth muscle cells.

Quo Vadis, Podocyte Actin Cytoskeleton?

Besides actin's effects on cell dynamic, anchorage and morphology, the actin cytoskeleton might have many more functions in podocytes, as described in detail in other cell types. However the podocyte field seems to be lagging behind. Here are some of these proposed actin-related mechanisms.

First, SD and FP proteins need to communicate with the cell body, in particular with the nucleus to coordinate SD and FP function with gene expression. Although most FP changes are fast and transient and most likely do not involve changes in transcriptional activity, major impacts especially when persistent and damaging, require changes in gene expression. It is possible that podocytes use LIM proteins as signal mediators, which are actin-associated proteins that shuttle between the actin cytoskeleton and the nuclear compartment and alter gene expression [93]. Zyxin, CARP and WTIP are LIM proteins that have been already described in podocytes (table 1). In fact, WTIP has been shown to travel along MT in podocyte primary processes [10] and it is likely that it can switch between actin and MT, dependent on the podocyte domain.

Second, actin fibers can form rails for the transport of molecular complexes including RNAs and proteins, as well as vesicles and even organelles. However, whether podocytes use their stress fibers as such transportation routes is still unknown. Based on the distance between the FP tip and the cell body, it is very likely that at least macromolecular complexes and mRNAs are transported, and that in the FP translational processes occur at local ribosomes, as described in neurons [94]. It is likely that as in neurons, the actin mRNA itself is transported via this mechanism, where zipcode-binding protein 1 (ZBP1) binds a specific nucleotide motif (ACACCC) in the actin 3′-UTR [95] and delivers the mRNA to the tip of the growing neurite. It is possible that this mechanism for local actin synthesis in the FP is also involved in FP formation during development and FP rearrangement in adult podocytes. It would be interesting to determine if mRNAs of other FP proteins also carry a zip-code motif and might undergo similar local translation.

Third, actin filaments regulate endo- and exocytosis, and recent work has led to the identification of proteins in podocytes that are involved in endocytotic processes in other cell types, like neurons [96, 97]. It is likely that podocytes have endocytotic activity for the release and/or uptake of soluble factors, or more likely to regulate the amount of transmembrane proteins like signaling receptors at the cell surface and thereby signal transduction and podocyte responsiveness. Myo1e-mediated regulation of surface expression of nephrin and Neph1 is a first indicator for the existence of such a mechanism [89]. Recent studies indicate the importance of endocytotic processes for proper podocyte homeostasis [98, 99], but also show that dependent on the cargo, podocyte endocytosis might induce podocyte injury and death [100].

Fourth, actin drives cytokinesis and in post-mitotic cells like podocytes this process must be silenced. Since under specific pathologic conditions as in HIVAN [101], podocytes can proliferate, cytokinetic properties are not switched off irreversibly. In many other pathological conditions, podocytes undergo mitosis without consequent cell division leading to pluripotency [102]. It would be important to understand why in these scenarios cytokinesis cannot be reactivated as this situations might be originally beneficial [9], but if persistent might lead to 'mitotic catastrophe' which has been described as a critical pathomechanism in different types of glomerular injury [102]. Following this thought, it is possible that podocytes adopted a post-mitotic state since they have a sophisticated cell morphology and actin cytoskeleton, and it would be too energy- and time-consuming to reorganize these structures and use them to form an actin-myosin contractile ring for cytokinesis followed by its rearrangement after completion of cell division. Or is it the other way around, that because podocytes stopped proliferating they were able to adapt such a sophisticated morphology?

When comparing podocyte features with other cell types in the body, striking similarities to neurons become apparent [56]. Like podocytes, neurons are also post-mitotic and develop a sophisticated cell morphology that contains MT- and actin-based protrusions, and actin-mediated cell plasticity is key for cell and overall tissue function. Of no surprise, podocyte biologists have started to look closely into neuron function and the complex mechanisms underlying actin dynamics in neuronal plasticity, and have identified novel proteins in podocytes which have been originally thought to be specific for neurons [97, 103, 104].

Conclusions and Outlook

About 25 years have passed between the identification of the contractile actin apparatus in podocyte FP and the visualization of podocyte movement in living animals [5, 21]. During this time period many actin-associated proteins have been identified in podocytes; some have been causatively linked to human glomerular disorders, while others have been shown to cause proteinuria when genetically modified in animals. Clearly, available large-scale expression analyses and '-omics' studies, and screening through their extensive lists of altered genes and proteins, will help to identify even more actin-associated proteins in podocytes. Furthermore, novel animal models like zebrafish where genetic modifications can be easily introduced and phenotypic readouts for glomerular changes, even if occurring only transiently, can be quantitatively studied, will help to determine if these proteins are important for podocyte function. Future studies are necessary to tackle some of the still unanswered important questions in the podocyte field: (a) Do podocytes under physiologic conditions move at all? (b) Do injured podocytes exhibit changes in cell motility, as suggested by many in vitro studies, and if so, do these alterations contribute to the injury process?

Although actin-related podocyte research will continue to unveil novel findings, the question arises if they will 'only' fill the remaining gaps in the mechanistic picture, or whether they will actually help to identify novel drug targets that can be used to specifically tackle glomerular diseases like MCD or FSGS. Most of the described actin regulators are expressed in many different cell types, and major side effects following pharmacological modification would be expected. In the cancer field, where interference with actin dynamics and cell migration would be beneficial by blocking metastatic growth and migration, the majority of drugs do not tackle the actin cytoskeleton, but upstream signaling events [105]. In general, the majority of currently approved drugs target protein kinases and cell surface reports, and to date G-protein coupled receptors (GPCR) are the most successful and potent drug targets in pharmacology [106]. Therefore, if the goal is to identify novel treatable pathomechanisms in glomerular diseases, one might want to focus on studying GPCR signaling in podocytes.

This should be a fruitful approach, since compared to 100 actin-associated proteins, only about 25 GPCRs have been identified in podocytes so far ... and the human genome encodes for about 1,000 GPCRs.

Disclosure Statement

The author has no conflicts of interest to disclose.

References

1 Mundel P, Shankland SJ: Podocyte biology and response to injury. J Am Soc Nephrol 2002;13:3005–3015.

2 Faul C, Asanuma K, Yanagida-Asanuma E, Kim K, Mundel P: Actin up: regulation of podocyte structure and function by components of the actin cytoskeleton. Trends Cell Biol 2007;17:428–437.

3 Rodewald R, Karnovsky MJ: Porous substructure of the glomerular slit diaphragm in the rat and mouse. J Cell Biol 1974;60:423–433.

4 Vasmant D, Maurice M, Feldman G: Cytoskeleton ultrastructure of podocytes and glomerular endothelial cells in rat and man. Anat Rec 1984;210:17–24.

5 Drenckhahn D, Franke RP: Ultrastructural organization of contractile and cytoskeletal proteins in glomerular podocytes of chicken, rat, and man. Lab Invest 1988;59:673–682.

6 Stamenkovic I, Skalli O, Gabbiani G: Distribution of intermediate filament proteins in normal and diseased human glomeruli. Am J Pathol 1986;125:465–475.

7 Chen J, Boyle S, Zhao M, Su W, Takahashi K, Davis L, et al: Differential expression of the intermediate filament protein nestin during renal development and its localization in adult podocytes. J Am Soc Nephrol 2006;17:1283–1291.

8 Kobayashi N, Heid HW, Sakai T, Kriz W, Huber G, Mundel P: Molecular characterization reveals identity of microtubule-associated proteins MAP3 and MAP4. Biochem Biophys Res Commun 2000;268:306–309.

9 Kriz W, Shirato I, Nagata M, LeHir M, Lemley KV: The podocyte's response to stress: the enigma of foot process effacement. Am J Physiol Renal Physiol 2013;304:F333–F347.

10 Kim JH, Konieczkowski M, Mukherjee A, Schechtman S, Khan S, Schelling JR, et al: Podocyte injury induces nuclear translocation of WTIP via microtubule-dependent transport. J Biol Chem 2010;285:9995–10004.

11 Wen D, You L, Zhang Q, Zhang L, Gu Y, Hao CM, et al: Upregulation of nestin protects podocytes from apoptosis induced by puromycin aminonucleoside. Am J Nephrol 2011;34:423–434.

12 Mouawad F, Aoudjit L, Jiang R, Szaszi K, Takano T: Role of guanine nucleotide exchange factor-H1 in complement-mediated RhoA activation in glomerular epithelial cells. J Biol Chem 2013, Epub ahead of print.

13 Saleem MA: New developments in steroid-resistant nephrotic syndrome. Pediatr Nephrol 2013;28:699–709.

14 Gbadegesin RA, Winn MP, Smoyer WE: Genetic testing in nephrotic syndrome – challenges and opportunities. Nat Rev Nephrol 2013;9:179–184.

15 Kriz W, LeHir M: Pathways to nephron loss starting from glomerular diseases – insights from animal models. Kidney Int 2005;67:404–419.

16 Oh J, Reiser J, Mundel P: Dynamic (re)organization of the podocyte actin cytoskeleton in the nephrotic syndrome. Pediatr Nephrol 2004;19:130–137.

17 Wharram BL, Goyal M, Wiggins JE, Sanden SK, Hussain S, Filipiak WE, et al: Podocyte depletion causes glomerulosclerosis: diphtheria toxin-induced podocyte depletion in rats expressing human diphtheria toxin receptor transgene. J Am Soc Nephrol 2005;16:2941–2952.

64 Gupta IR, Baldwin C, Auguste D, Ha KC, El Andalousi J, Fahiminiya S, et al: ARHGDIA: a novel gene implicated in nephrotic syndrome. J Med Genet 2013;50:330–338.
65 Shibata S, Nagase M, Yoshida S, Kawarazaki W, Kurihara H, Tanaka H, et al: Modification of mineralocorticoid receptor function by Rac1 GTPase: implication in proteinuric kidney disease. Nat Med 2008;14:1370–1376.
66 Lu TC, He JC, Wang ZH, Feng X, Fukumi-Tominaga T, Chen N, et al: HIV-1 Nef disrupts the podocyte actin cytoskeleton by interacting with diaphanous interacting protein. J Biol Chem 2008;283:8173–8182.
67 Kaplan JM, Kim SH, North KN, Rennke H, Correia LA, Tong HQ, et al: Mutations in ACTN4, encoding α-actinin-4, cause familial focal segmental glomerulosclerosis. Nat Genet 2000;24:251–256.
68 Weins A, Kenlan P, Herbert S, Le TC, Villegas I, Kaplan BS, et al: Mutational and biological analysis of α-actinin-4 in focal segmental glomerulosclerosis. J Am Soc Nephrol 2005;16:3694–3701.
69 Ward SM, Weins A, Pollak MR, Weitz DA: Dynamic viscoelasticity of actin cross-linked with wild-type and disease-causing mutant α-actinin-4. Biophys J 2008;95:4915–4923.
70 Pawson T, Scott JD: Signaling through scaffold, anchoring, and adaptor proteins. Science 1997;278:2075–2080.
71 Schell C, Baumhakl L, Salou S, Conzelmann AC, Meyer C, Helmstadter M, et al: N-wasp is required for stabilization of podocyte foot processes. J Am Soc Nephrol 2013;24:713–721.
72 Van Duijn TJ, Anthony EC, Hensbergen PJ, Deelder AM, Hordijk PL: Rac1 recruits the adapter protein CMS/CD2AP to cell-cell contacts. J Biol Chem 2010;285:20137–20146.
73 Tossidou I, Kardinal C, Peters I, Kriz W, Shaw A, Dikic I, et al: CD2AP/CIN85 balance determines receptor tyrosine kinase signaling response in podocytes. J Biol Chem 2007;282: 7457–7464.
74 Huber TB, Hartleben B, Kim J, Schmidts M, Schermer B, Keil A, et al: Nephrin and CD2AP associate with phosphoinositide 3-OH kinase and stimulate AKT-dependent signaling. Mol Cell Biol 2003;23:4917–4928.
75 Yaddanapudi S, Altintas MM, Kistler AD, Fernandez I, Moller CC, Wei C, et al: CD2AP in mouse and human podocytes controls a proteolytic program that regulates cytoskeletal structure and cellular survival. J Clin Invest 2011;121:3965–3980.
76 Katoh K, Kano Y, Noda Y: Rho-associated kinase-dependent contraction of stress fibres and the organization of focal adhesions. J R Soc Interface 2011;8:305–311.
77 White GE, Gimbrone MA Jr, Fujiwara K: Factors influencing the expression of stress fibers in vascular endothelial cells in situ. J Cell Biol 1983;97:416–424.
78 Wong JS, Iorns E, Rheault MN, Ward TM, Rashmi P, Weber U, et al: Rescue of tropomyosin deficiency in *Drosophila* and human cancer cells by synaptopodin reveals a role of tropomyosin-α in RhoA stabilization. EMBO J 2012;31:1028–1040.
79 Kamm KE, Stull JT: The function of myosin and myosin light chain kinase phosphorylation in smooth muscle. Annu Rev Pharmacol Toxicol 1985;25:593–620.
80 Kawano Y, Yoshimura T, Kaibuchi K: Smooth muscle contraction by small GTPase Rho. Nagoya J Med Sci 2002;65:1–8.
81 Nijenhuis T, Sloan AJ, Hoenderop JG, Flesche J, van Goor H, Kistler AD, et al: Angiotensin II contributes to podocyte injury by increasing TRPC6 expression via an NFAT-mediated positive feedback signaling pathway. Am J Pathol 2011;179:1719–1732.
82 Arrondel C, Vodovar N, Knebelmann B, Grunfeld JP, Gubler MC, Antignac C, et al: Expression of the non-muscle myosin heavy chain IIA in the human kidney and screening for MYH9 mutations in Epstein and Fechtner syndromes. J Am Soc Nephrol 2002;13:65–74.
83 Johnstone DB, Zhang J, George B, Leon C, Gachet C, Wong H, et al: Podocyte-specific deletion of Myh9 encoding non-muscle myosin heavy chain 2A predisposes mice to glomerulopathy. Mol Cell Biol 2011;31:2162–2170.
84 Mele C, Iatropoulos P, Donadelli R, Calabria A, Maranta R, Cassis P, et al: MYO1E mutations and childhood familial focal segmental glomerulosclerosis. N Engl J Med 2011;365: 295–306.

85 Sanna-Cherchi S, Burgess KE, Nees SN, Caridi G, Weng PL, Dagnino M, et al: Exome sequencing identified MYO1E and NEIL1 as candidate genes for human autosomal recessive steroid-resistant nephrotic syndrome. Kidney Int 2011;80:389–396.
86 Krendel M, Kim SV, Willinger T, Wang T, Kashgarian M, Flavell RA, et al: Disruption of myosin 1e promotes podocyte injury. J Am Soc Nephrol 2009;20:86–94.
87 Chase SE, Encina CV, Stolzenburg LR, Tatum AH, Holzman LB, Krendel M: Podocyte-specific knockout of myosin 1e disrupts glomerular filtration. Am J Physiol Renal Physiol 2012;303:F1099–F1106.
88 Arif E, Kumari B, Wagner MC, Zhou W, Holzman LB, Nihalani D: Myo1c is an unconventional myosin required for zebrafish glomerular development. Kidney Int 2013;84: 1154–1165.
89 Arif E, Wagner MC, Johnstone DB, Wong HN, George B, Pruthi PA, et al: Motor protein Myo1c is a podocyte protein that facilitates the transport of slit diaphragm protein Neph1 to the podocyte membrane. Mol Cell Biol 2011;31:2134–2150.
90 Jin X, Wang W, Mao J, Shen H, Fu H, Wang X, et al: Overexpression of Myo1e in mouse podocytes enhances cellular endocytosis, migration, and adhesion. J Cell Biochem 2014; 115:410–419.
91 Saleem MA, Zavadil J, Bailly M, McGee K, Witherden IR, Pavenstadt H, et al: The molecular and functional phenotype of glomerular podocytes reveals key features of contractile smooth muscle cells. Am J Physiol Renal Physiol 2008;295:F959–F970.
92 Wennmann DO, Hsu HH, Pavenstaedt H: The renin-angiotensin-aldosterone system in podocytes. Seminar Nephrol 2012;32:377–384.
93 Kadrmas JL, Beckerle MC: The LIM domain: from the cytoskeleton to the nucleus. Nat Rev Mol Cell Biol 2004;5:920–931.
94 Koenig E: Organized ribosome-containing structural domains in axons. Results Probl Cell Differ 2009;48:173–191.
95 Ross AF, Oleynikov Y, Kislauskis EH, Taneja KL, Singer RH: Characterization of a beta-actin mRNA zipcode-binding protein. Mol Cell Biol 1997;17:2158–2165.
96 Soda K, Balkin DM, Ferguson SM, Paradise S, Milosevic I, Giovedi S, et al: Role of dynamin, synaptojanin, and endophilin in podocyte foot processes. J Clin Invest 2012;122:4401–4411.
97 Rastaldi MP, Armelloni S, Berra S, Calvaresi N, Corbelli A, Giardino LA, et al: Glomerular podocytes contain neuron-like functional synaptic vesicles. FASEB J 2006;20:976–978.
98 Bechtel W, Helmstadter M, Balica J, Hartleben B, Kiefer B, Hrnjic F, et al: Vps34 deficiency reveals the importance of endocytosis for podocyte homeostasis. J Am Soc Nephrol 2013;24:727–743.
99 Chen J, Chen MX, Fogo AB, Harris RC, Chen JK: mVps34 deletion in podocytes causes glomerulosclerosis by disrupting intracellular vesicle trafficking. J Am Soc Nephrol 2013;24:198–207.
100 Okamura K, Dummer P, Kopp J, Qiu L, Levi M, Faubel S, et al: Endocytosis of albumin by podocytes elicits an inflammatory response and induces apoptotic cell death. PLoS One 2013;8:e54817.
101 Barisoni L, Kriz W, Mundel P, D'Agati V: The dysregulated podocyte phenotype: a novel concept in the pathogenesis of collapsing idiopathic focal segmental glomerulosclerosis and HIV-associated nephropathy. J Am Soc Nephrol 1999;10:51–61.
102 Liapis H, Romagnani P, Anders HJ: New insights into the pathology of podocyte loss: mitotic catastrophe. Am J Pathol 2013;183:1364–1374.
103 Giardino L, Armelloni S, Corbelli A, Mattinzoli D, Zennaro C, Guerrot D, et al: Podocyte glutamatergic signaling contributes to the function of the glomerular filtration barrier. J Am Soc Nephrol 2009;20: 1929–1940.
104 Sistani L, Rodriguez PQ, Hultenby K, Uhlen M, Betsholtz C, Jalanko H, et al: Neuronal proteins are novel components of podocyte major processes and their expression in glomerular crescents supports their role in crescent formation. Kidney Int 2013;83:63–71.
105 Zhang J, Yang PL, Gray NS: Targeting cancer with small molecule kinase inhibitors. Nat Rev Cancer 2009;9:28–39.
106 Abrol RG, Goddard WA: G protein-coupled receptors: conformational 'gatekeepers' of transmembrane signal transduction and diversification. R Soc Chem Extracell Intracell Signal 2011;10:188–229.

107 Lehtonen S, Ryan JJ, Kudlicka K, Iino N, Zhou H, Farquhar MG: Cell junction-associated proteins IQGAP1, MAGI-2, CASK, spectrins, and α-actinin are components of the nephrin multiprotein complex. Proc Natl Acad Sci USA 2005;102:9814–9819.
108 Welsch T, Endlich N, Kriz W, Endlich K: CD2AP and p130Cas localize to different F-actin structures in podocytes. Am J Physiol Renal Physiol 2001;281:F769–F777.
109 Ichimura K, Kurihara H, Sakai T: Actin filament organization of foot processes in rat podocytes. J Histochem Cytochem 2003;51: 1589–1600.
110 Trenchev P, Dorling J, Webb J, Holborow EJ: Localization of smooth muscle-like contractile proteins in kidney by immunoelectron microscopy. J Anat 1976;121:85–95.
111 Lachapelle M, Bendayan M: Contractile proteins in podocytes: immunocytochemical localization of actin and α-actinin in normal and nephrotic rat kidneys. Virchows Archiv B Cell Pathol 1991;60:105–111.
112 Smoyer WE, Mundel P, Gupta A, Welsh MJ: Podocyte α-actinin induction precedes foot process effacement in experimental nephrotic syndrome. Am J Physiol 1997; 273:F150–F157.
113 Ferrandi M, Cusi D, Molinari I, Del Vecchio L, Barlassina C, Rastaldi MP, et al: Alpha- and beta-adducin polymorphisms affect podocyte proteins and proteinuria in rodents and decline of renal function in human IgA nephropathy. J Mol Med (Berl) 2010;88:203–217.
114 Reiser J, Kriz W, Kretzler M, Mundel P: The glomerular slit diaphragm is a modified adherens junction. J Am Soc Nephrol 2000;11: 1–8.
115 Yang Y, Guo L, Blattner SM, Mundel P, Kretzler M, Wu C: Formation and phosphorylation of the PINCH-1-integrin linked kinase-α-parvin complex are important for regulation of renal glomerular podocyte adhesion, architecture, and survival. J Am Soc Nephrol 2005;16:1966–1976.
116 Burghardt T, Kastner J, Suleiman H, Rivera-Milla E, Stepanova N, Lottaz C, et al: LMX1B is essential for the maintenance of differentiated podocytes in adult kidneys. J Am Soc Nephrol 2013;24:1830–1848.
117 Xu X, Patrakka J, Sistani L, Uhlen M, Jalanko H, Betsholtz C, et al: Expression of novel podocyte-associated proteins sult1b1 and ankrd25. Nephron Exp Nephrol 2011; 117:e39–e46.
118 Hartleben B, Schweizer H, Lubben P, Bartram MP, Moller CC, Herr R, et al: Neph-nephrin proteins bind the Par3-Par6-atypical protein kinase C (aPKC) complex to regulate podocyte cell polarity. J Biol Chem 2008;283:23033–23038.
119 Huber TB, Hartleben B, Winkelmann K, Schneider L, Becker JU, Leitges M, et al: Loss of podocyte aPKCλ/iota causes polarity defects and nephrotic syndrome. J Am Soc Nephrol 2009;20:798–806.
120 Hirose T, Satoh D, Kurihara H, Kusaka C, Hirose H, Akimoto K, et al: An essential role of the universal polarity protein, aPKCλ, on the maintenance of podocyte slit diaphragms. PLoS One 2009;4:e4194.
121 Hsu YH, Lin WL, Hou YT, Pu YS, Shun CT, Chen CL, et al: Podocalyxin EBP50 ezrin molecular complex enhances the metastatic potential of renal cell carcinoma through recruiting Rac1 guanine nucleotide exchange factor ARHGEF7. Am J Pathol 2010; 176:3050–3061.
122 Kim JH, Mukherjee A, Madhavan SM, Konieczkowski M, Sedor JR: WT1-interacting protein (Wtip) regulates podocyte phenotype by cell-cell and cell-matrix contact reorganization. Am J Physiol Renal Physiol 2012;302:F103–F115.
123 Miao J, Fan Q, Cui Q, Zhang H, Chen L, Wang S, et al: Newly identified cytoskeletal components are associated with dynamic changes of podocyte foot processes. Nephrol Dial Transplant 2009;24:3297–3305.
124 Dai C, Stolz DB, Kiss LP, Monga SP, Holzman LB, Liu Y: Wnt/β-catenin signaling promotes podocyte dysfunction and albuminuria. J Am Soc Nephrol 2009;20:1997–2008.
125 Heikkila E, Ristola M, Endlich K, Lehtonen S, Lassila M, Havana M, et al: Densin and β-catenin form a complex and co-localize in cultured podocyte cell junctions. Mol Cell Biochem 2007;305:9–18.
126 Kato H, Gruenwald A, Suh JH, Miner JH, Barisoni-Thomas L, Taketo MM, et al: Wnt/β-catenin pathway in podocytes integrates cell adhesion, differentiation, and survival. J Biol Chem 2011;286:26003–26015.

127 Matsuura K, Uesugi N, Hijiya N, Uchida T, Moriyama M: Upregulated expression of cardiac ankyrin-repeated protein in renal podocytes is associated with proteinuria severity in lupus nephritis. Hum Pathol 2007;38:410–419.
128 Fukasawa H, Bornheimer S, Kudlicka K, Farquhar MG: Slit diaphragms contain tight junction proteins. J Am Soc Nephrol 2009;20:1491–1503.
129 Lehtonen S, Lehtonen E, Kudlicka K, Holthofer H, Farquhar MG: Nephrin forms a complex with adherens junction proteins and CASK in podocytes and in Madin-Darby canine kidney cells expressing nephrin. Am J Pathol 2004;165:923–936.
130 Shih NY, Li J, Karpitskii V, Nguyen A, Dustin ML, Kanagawa O, et al: Congenital nephrotic syndrome in mice lacking CD2-associated protein. Science 1999;286:312–315.
131 Lehtonen S, Zhao F, Lehtonen E: CD2-associated protein directly interacts with the actin cytoskeleton. Am J Physiol Renal Physiol 2002;283:F734–F743.
132 Garg P, Verma R, Cook L, Soofi A, Venkatareddy M, George B, et al: Actin-depolymerizing factor cofilin-1 is necessary in maintaining mature podocyte architecture. J Biol Chem 2010;285:22676–22688.
133 Ashworth S, Teng B, Kaufeld J, Miller E, Tossidou I, Englert C, et al: Cofilin-1 inactivation leads to proteinuria – studies in zebrafish, mice and humans. PLoS One 2010;5:e12626.
134 Kobayashi T, Notoya M, Shinosaki T, Kurihara H: Cortactin interacts with podocalyxin and mediates morphological change of podocytes through its phosphorylation. Nephron Exp Nephrol 2009;113:e89–e96.
135 Zhao J, Bruck S, Cemerski S, Zhang L, Butler B, Dani A, et al: CD2AP links cortactin and capping protein at the cell periphery to facilitate formation of lamellipodia. Mol Cell Biol 2013;33:38–47.
136 George B, Verma R, Soofi AA, Garg P, Zhang J, Park TJ, et al: Crk1/2-dependent signaling is necessary for podocyte foot process spreading in mouse models of glomerular disease. J Clin Invest 2012;122:674–692.
137 Worthmann K, Leitges M, Teng B, Sestu M, Tossidou I, Samson T, et al: Def-6, a novel regulator of small GTPases in podocytes, acts downstream of atypical protein kinase C (aPKC) λ/iota. Am J Pathol 2013;183:1945–1959.
138 Ahola H, Heikkila E, Astrom E, Inagaki M, Izawa I, Pavenstadt H, et al: A novel protein, densin, expressed by glomerular podocytes. J Am Soc Nephrol 2003;14:1731–1747.
139 Laurin M, Dumouchel A, Fukui Y, Cote JF: The Rac-specific exchange factors Dock1 and Dock5 are dispensable for the establishment of the glomerular filtration barrier in vivo. Small GTPases 2013;4.
140 Peitsch WK, Hofmann I, Endlich N, Pratzel S, Kuhn C, Spring H, et al: Cell biological and biochemical characterization of drebrin complexes in mesangial cells and podocytes of renal glomeruli. J Am Soc Nephrol 2003;14:1452–1463.
141 Sever S, Altintas MM, Nankoe SR, Moller CC, Ko D, Wei C, et al: Proteolytic processing of dynamin by cytoplasmic cathepsin L is a mechanism for proteinuric kidney disease. J Clin Invest 2007;117:2095–2104.
142 Gu C, Yaddanapudi S, Weins A, Osborn T, Reiser J, Pollak M, et al: Direct dynamin-actin interactions regulate the actin cytoskeleton. EMBO J 2010;29:3593–3606.
143 Orlando RA, Takeda T, Zak B, Schmieder S, Benoit VM, McQuistan T, et al: The glomerular epithelial cell anti-adhesin podocalyxin associates with the actin cytoskeleton through interactions with ezrin. J Am Soc Nephrol 2001;12:1589–1598.
144 Fukasawa H, Obayashi H, Schmieder S, Lee J, Ghosh P, Farquhar MG: Phosphorylation of podocalyxin (Ser415) prevents RhoA and ezrin activation and disrupts its interaction with the actin cytoskeleton. Am J Pathol 2011;179:2254–2265.
145 Morigi M, Buelli S, Angioletti S, Zanchi C, Longaretti L, Zoja C, et al: In response to protein load podocytes reorganize cytoskeleton and modulate endothelin-1 gene: implication for permselective dysfunction of chronic nephropathies. Am J Pathol 2005;166:1309–1320.
146 Ma H, Togawa A, Soda K, Zhang J, Lee S, Ma M, et al: Inhibition of podocyte FAK protects against proteinuria and foot process effacement. J Am Soc Nephrol 2010;21:1145–1156.
147 Venkatareddy M, Cook L, Abuarquob K, Verma R, Garg P: Nephrin regulates lamellipodia formation by assembling a protein complex that includes Ship2, filamin and lamellipodin. PLoS One 2011;6:e28710.

148 Tan R, Patni H, Tandon P, Luan L, Sharma B, Salhan D, et al: Nef interaction with actin compromises human podocyte actin cytoskeletal integrity. Exp Mol Pathol 2013;94:51–57.
149 Brown EJ, Schlondorff JS, Becker DJ, Tsukaguchi H, Tonna SJ, Uscinski AL, et al: Mutations in the formin gene INF2 cause focal segmental glomerulosclerosis. Nat Genet 2010;42:72–76.
150 Li Y, Kang YS, Dai C, Kiss LP, Wen X, Liu Y: Epithelial-to-mesenchymal transition is a potential pathway leading to podocyte dysfunction and proteinuria. Am J Pathol 2008;172:299–308.
151 Samejima K, Nakatani K, Suzuki D, Asai O, Sakan H, Yoshimoto S, et al: Clinical significance of fibroblast-specific protein-1 expression on podocytes in patients with focal segmental glomerulosclerosis. Nephron Clin Pract 2012;120:c1–c7.
152 Harvey SJ, Jarad G, Cunningham J, Goldberg S, Schermer B, Harfe BD, et al: Podocyte-specific deletion of dicer alters cytoskeletal dynamics and causes glomerular disease. J Am Soc Nephrol 2008;19:2150–2158.
153 Perisic L, Lal M, Hulkko J, Hultenby K, Onfelt B, Sun Y, et al: Plekhh2, a novel podocyte protein downregulated in human focal segmental glomerulosclerosis, is involved in matrix adhesion and actin dynamics. Kidney Int 2012;82:1071–1083.
154 Guo L, Sanders PW, Woods A, Wu C: The distribution and regulation of integrin-linked kinase in normal and diabetic kidneys. Am J Pathol 2001;159:1735–1742.
155 Rigothier C, Auguste P, Welsh GI, Lepreux S, Deminiere C, Mathieson PW, et al: IQGAP1 interacts with components of the slit diaphragm complex in podocytes and is involved in podocyte migration and permeability in vitro. PLoS One 2012;7:e37695.
156 Duning K, Schurek EM, Schluter M, Bayer M, Reinhardt HC, Schwab A, et al: KIBRA modulates directional migration of podocytes. J Am Soc Nephrol 2008;19:1891–1903.
157 Qu H, Tu Y, Shi X, Larjava H, Saleem MA, Shattil SJ, et al: Kindlin-2 regulates podocyte adhesion and fibronectin matrix deposition through interactions with phosphoinositides and integrins. J Cell Sci 2011;124:879–891.
158 Sun H, Schlondorff JS, Brown EJ, Higgs HN, Pollak MR: Rho activation of mDia formins is modulated by an interaction with inverted formin 2 (INF2). Proc Natl Acad Sci USA 2011;108:2933–2938.
159 Sun H, Schlondorff J, Higgs HN, Pollak MR: Inverted formin 2 regulates actin dynamics by antagonizing Rho/diaphanous-related formin signaling. J Am Soc Nephrol 2013;24:917–929.
160 Patrie KM, Drescher AJ, Goyal M, Wiggins RC, Margolis B: The membrane-associated guanylate kinase protein MAGI-1 binds megalin and is present in glomerular podocytes. J Am Soc Nephrol 2001;12:667–677.
161 Srichai MB, Konieczkowski M, Padiyar A, Konieczkowski DJ, Mukherjee A, Hayden PS, et al: A WT1 co-regulator controls podocyte phenotype by shuttling between adhesion structures and nucleus. J Biol Chem 2004;279:14398–14408.
162 Jensen AM, Walker C, Westerfield M: mosaic eyes: a zebrafish gene required in pigmented epithelium for apical localization of retinal cell division and lamination. Development 2001;128:95–105.
163 Kramer-Zucker AG, Wiessner S, Jensen AM, Drummond IA: Organization of the pronephric filtration apparatus in zebrafish requires nephrin, podocin and the FERM domain protein mosaic eyes. Dev Biol 2005;285:316–329.
164 Bi J, Chase SE, Pellenz CD, Kurihara H, Fanning AS, Krendel M: Myosin 1e is a component of the glomerular slit-diaphragm complex that regulates actin reorganization during cell-cell contact formation in podocytes. Am J Physiol Renal Physiol 2013;305:F532–F544.
165 Zhang SY, Kamal M, Dahan K, Pawlak A, Ory V, Desvaux D, et al: c-mip impairs podocyte proximal signaling and induces heavy proteinuria. Sci Signal 2010;3:ra39.
166 Jones N, Blasutig IM, Eremina V, Ruston JM, Bladt F, Li H, et al: Nck adaptor proteins link nephrin to the actin cytoskeleton of kidney podocytes. Nature 2006;440:818–823.
167 Verma R, Kovari I, Soofi A, Nihalani D, Patrie K, Holzman LB: Nephrin ectodomain engagement results in Src kinase activation, nephrin phosphorylation, Nck recruitment, and actin polymerization. J Clin Invest 2006;116:1346–1359.

168 Blasutig IM, New LA, Thanabalasuriar A, Dayarathna TK, Goudreault M, Quaggin SE, et al: Phosphorylated YDXV motifs and Nck SH2/SH3 adaptors act cooperatively to induce actin reorganization. Mol Cell Biol 2008;28:2035–2046.

169 He JC, Husain M, Sunamoto M, D'Agati VD, Klotman ME, Iyengar R, et al: Nef stimulates proliferation of glomerular podocytes through activation of Src-dependent Stat3 and MAPK1,2 pathways. J Clin Invest 2004;114:643–651.

170 Takeda T, McQuistan T, Orlando RA, Farquhar MG: Loss of glomerular foot processes is associated with uncoupling of podocalyxin from the actin cytoskeleton. J Clin Invest 2001;108:289–301.

171 Schmieder S, Nagai M, Orlando RA, Takeda T, Farquhar MG: Podocalyxin activates RhoA and induces actin reorganization through NHERF1 and ezrin in MDCK cells. J Am Soc Nephrol 2004;15:2289–2298.

172 Togashi M, Okuyama S, Wakui H, Komatsuda A, Suzuki D, Toyoda M, et al: Interaction of α-actinin-4 with class I PxxP motif-containing OK/SW-CL.16 protein. Nephron Exp Nephrol 2007;107:e65–e72.

173 Usui J, Kurihara H, Shu Y, Tomari S, Kanemoto K, Koyama A, et al: Localization of intercellular adherens junction protein p120 catenin during podocyte differentiation. Anat Embryol (Berl) 2003;206:175–184.

174 Bains R, Furness PN, Critchley DR: A quantitative immunofluorescence study of glomerular cell adhesion proteins in proteinuric states. J Pathol 1997;183:272–280.

175 Endlich N, Schordan E, Cohen CD, Kretzler M, Lewko B, Welsch T, et al: Palladin is a dynamic actin-associated protein in podocytes. Kidney Int 2009;75:214–226.

176 Topham PS, Haydar SA, Kuphal R, Lightfoot JD, Salant DJ: Complement-mediated injury reversibly disrupts glomerular epithelial cell actin microfilaments and focal adhesions. Kidney Int 1999;55:1763–1775.

177 Kerjaschki D, Ojha PP, Susani M, Horvat R, Binder S, Hovorka A, et al: A β_1-integrin receptor for fibronectin in human kidney glomeruli. Am J Pathol 1989;134:481–489.

178 Sistani L, Duner F, Udumala S, Hultenby K, Uhlen M, Betsholtz C, et al: Pdlim2 is a novel actin-regulating protein of podocyte foot processes. Kidney Int 2011;80:1045–1054.

179 Shi X, Qu H, Kretzler M, Wu C: Roles of PINCH-2 in regulation of glomerular cell shape change and fibronectin matrix deposition. Am J Physiol Renal Physiol 2008; 295:F253–F263.

180 Harita Y, Kurihara H, Kosako H, Tezuka T, Sekine T, Igarashi T, et al: Phosphorylation of nephrin triggers Ca^{2+} signaling by recruitment and activation of phospholipase Cγ1. J Biol Chem 2009;284:8951–8962.

181 Patrakka J, Xiao Z, Nukui M, Takemoto M, He L, Oddsson A, et al: Expression and subcellular distribution of novel glomerulus-associated proteins dendrin, Ehd3, Sh2d4a, Plekhh2, and 2310066E14Rik. J Am Soc Nephrol 2007;18:689–697.

182 Hoshi S, Nomoto K, Kuromitsu J, Tomari S, Nagata M: High glucose induced VEGF expression via PKC and ERK in glomerular podocytes. Biochem Biophys Res Commun 2002;290:177–184.

183 Tossidou I, Teng B, Menne J, Shushakova N, Park JK, Becker JU, et al: Podocytic PKC-α is regulated in murine and human diabetes and mediates nephrin endocytosis. PLoS One 2010;5:e10185.

184 Li M, Armelloni S, Ikehata M, Corbelli A, Pesaresi M, Calvaresi N, et al: Nephrin expression in adult rodent central nervous system and its interaction with glutamate receptors. J Pathol 2011;225:118–128.

185 Hartleben B, Widmeier E, Wanner N, Schmidts M, Kim ST, Schneider L, et al: Role of the polarity protein Scribble for podocyte differentiation and maintenance. PLoS One 2012;7:e36705.

186 Wasik AA, Polianskyte-Prause Z, Dong MQ, Shaw AS, Yates JR 3rd, Farquhar MG, et al: Septin 7 forms a complex with CD2AP and nephrin and regulates glucose transporter trafficking. Mol Biol Cell 2012;23: 3370–3379.

187 Inomata S, Sakatsume M, Sakamaki Y, Wang X, Goto S, Yamamoto T, et al: Expression of SM22α (transgelin) in glomerular and interstitial renal injury. Nephron Exp Nephrol 2011;117:e104–e113.

188 Marshall CB, Krofft RD, Blonski MJ, Kowalewska J, Logar CM, Pippin JW, et al: Role of smooth muscle protein SM22α in glomerular epithelial cell injury. Am J Physiol Renal Physiol 2011;300:F1026–F1042.

filtrate are activated by ammonia at the tubular epithelial brush border, and subsequently induce leukocyte infiltration to the renal interstitium [8]. In addition, the complement membrane attack complex activates TEC to enter a pro-inflammatory state, in which TEC release arachidonic acid products and pro-inflammatory interleukin (IL)-6, IL-1β and tumor necrosis factor-α (TNF-α) [9]. In vitro studies have demonstrated that albumin induces the release of TNF-α, IL-8, monocyte chemotactic protein-1 (MCP-1) and platelet-derived growth factor (PDGF) by TEC [10, 11]. This serves as a trigger of interstitial inflammation and fibrosis. Minimum impairment of podocytic function may cause microalbuminuria, where serum proteins or growth factors may leak into the urinary space and activate the TEC, resulting in increased synthesis of cytokines and chemokines [12]. The deleterious impact of filtered proteins on the TEC may be facilitated by additional metabolic factors. For example, high glucose enhances matrix metalloproteinase (MMP) activity on TEC in DN, and leads to the shedding of the megalin/cubilin complex, which is responsible for impaired tubular clearance of filtered proteins [13].

In tubulointerstitial injury associated with podocytic damage, cytokines and growth factors originating from podocytes and infiltrating cells play an indisputable role. Upon injury, podocytes release various cytokines including TNF-α, vascular endothelial growth factor (VEGF), transforming growth factor-β (TGF-β), granulocyte-macrophage colony-stimulating factor (GM-CSF) and IL-6 [14]. These cytokines may reach TEC through the tubule lumen, via delivery through peritubular capillaries or by diffusion through the interstitial space, and can be detected in the urine from patients and experimental animals with podocytic injury. The pleiotropic active TGF-β proteins could be secreted by the podocytes, as well as the TEC [15]. A murine model with transgenic expression of active TGF-β receptor type 1 in TEC has demonstrated that activation of the TGF-β signalling pathway alone induces dedifferentiation, apoptosis, necrosis and reactive oxygen species (ROS) generation in TEC, followed by the decline of renal function and inflammatory cell infiltration [16]. Podocytic injury also generates ROS and amplifies both the immune and non-immune glomerular injury [17]. Podocytes produce TGF-β to modulate extracellular matrix (ECM) protein production by TEC, and increase synthesis of ECM proteins including proteoglycans, fibronectin, type IV collagen, and laminin [18]. On the other hand, TGF-β may play a regulatory function on matrix synthesis through inhibition of the release of collagenases, or stimulation of the production of metalloproteinase inhibitors. In addition, high concentration of angiotensin II (AngII) and TGF-β may alter the TEC phenotype towards fibroblasts by epithelial-to-mesenchymal transition (EMT), which contributes to interstitial fibrosis and tubular atrophy [19].

Histological analyses of interstitial leukocyte infiltration in experimental glomerular diseases have suggested that tubulointerstitial inflammation begins in the urinary pole of glomeruli in which TEC are closest to the inflamed podocytes [20]. Podocytic cytokines may also lead to an upregulated expression of human leukocyte antigen (HLA) class II antigens and adhesion molecules by TEC, and trigger tubulointerstitial damage through leukocyte recruitment [21]. Infiltrating activated T cells and monocytes produce several inflammatory cytokines at sites of inflammation. These cytokines activate both the infiltrating inflammatory cells, as well as the resident TEC. Furthermore, through the increased expression of CD40L, activated T cells can also directly activate TEC [22]. This activation is further reinforced under the synergistic action of different cytokines.

Progressive activation of TEC leads to their apoptosis and results in nephron loss. Increased rate of oxygen consumption in the surviving nephrons increases ammonia and ROS production by the TEC [23]. ROS causes TEC damage directly in drug- or virus-induced tubular toxicity, or indirectly in CKD such as DN [24, 25]. High glucose stimulates the generation of ROS in TEC, and overexpression of catalase in diabetic mice reduces proximal tubular apoptosis, demonstrating the important role of oxidant injury in DN.

With the progressive amplification of inflammatory processes by the infiltrating inflammatory cells, significant loss of podocyte number occurs which causes misdirected filtration into the periglomerular interstitium, perpetuation of tubular destruction and the progression of kidney failure. In contrast, if there is no podocyte loss and protein leakage is confined across the filtration barrier, the damages may be reversible and the injured podocytes are repaired.

Will Injury to TEC Affect Podocyte?

Although glomerular or podocytic damage usually precede tubular injury, it is the latter that activates the irreversible tubulointerstitial fibrosis, thus leading to the progression to ESRD. Acute tubulointerstitial damage is frequently initiated by ischemia and nephrotoxins, and is associated with autoimmune disorders and infections. The mechanism of tubular damage is mostly immune-mediated, involving a marked interstitial infiltrate consisting of macrophages and T lymphocytes. A spontaneous recovery is often observed and the temporary tubular damage will less likely lead to the development of podocytopathy or glomerulopathy.

Chronic tubular interstitial nephritis is induced by a much broader spectrum of insults. Infiltration of leukocytes is commonly seen in acute tubulointerstitial nephritis, which later develops into tubular atrophy and interstitial fibrosis at advanced stages. Local activation of tubulointerstitial cells, as well as the infiltrating leukocytes, triggers the release of an array of inflammatory cytokines and growth factors, which ultimately promote fibrogenesis in the interstitium. The tubulointerstitial compartment may be damaged directly by different insults, or indirectly through a cascade of inflammatory or immune reactions. Direct tubulointerstitial injury is usually focal and the extent of damage depends on the duration of exposure to the insults and their nature.

Although most studies on acute kidney injury (AKI) have been focused on the pathophysiologic alteration in the tubulointerstitium, AKI is often associated with glomerular abnormalities. Local release of cytokines by inflammatory cells and TEC into the injured kidney initiates the development of inflammation in AKI. Ischemic AKI increases the expression of IL-2, IL-6, IL-10, interferon-γ (IFN-γ), TGF-β, GM-CSF, chemokine (C-X-C motif) ligand 1 (CXCL1), MCP-1, and macrophage inflammatory protein-2 (MIP-2) in the kidney [26]. Direct injury to tubules also increases circulating pro-inflammatory cytokines. In rat model of cisplatin nephrotoxicity, levels of TNF-α are increased in circulation, kidney, and urine [27]. Upregulated serum IL-6 has been demonstrated in rodent models of ischemic AKI [28]. In sepsis-induced AKI mice, there is elevated serum concentration of IL-6, TNF-α, MCP-1, and IL-10 [29]. Research into the potential effect of these increased circulating cytokines on podocytes during AKI is scarce. Podocytes express receptors for complements, TGF-β, AngII and IL-6, and if the glomerular endothelial cells are also damaged, these receptors will respond to the corresponding circulating cytokines or ligands in AKI and initiate podocytic injury. Podocytes have been shown to play a pathogenic role in septic AKI. Podocytes express IL-6 receptors and LPS induces IL-6 expression in podocytes, and thus provokes the pro-inflammatory response through an autocrine loop [30]. Logically predicted, the mediators released by the inflammatory cells and TEC during AKI may elicit the inflammatory and pro-apoptotic responses in podocytes. The effect of AKI on podocytes does not necessarily follow a one-way track. In septic burn patients affected by severe systemic inflammatory response, the associated proteinuria displays a mixed glomerular and tubular pattern, suggesting that there is a simultaneous decrease of tubular reabsorption and increase of glomerular permeability.

Insults that induce mild foot process effacement and abnormal slit diaphragm protein reorganization may render a reversible podocytic injury and repair. Injured podocytes are gradually recovered once the insults are removed and proteinuria normalized. However, if the podocytes are persistently activated, mediators released from the injured podocytes, particularly in the vicinity of urinary pole, exert cumulative effects on the TEC. These mediators include microalbuminuria, ROS, AngII, TGF-β, MMP and secreted protein acidic and

Table 1. Major molecular mediators secreted by activated TEC

Adhesion molecules	CD44, ICAM, integrins, VCAM
Cytokines	
Pro-inflammatory	IFN-γ, IL-1, IL-6, TNF-α
Pro-fibrotic	AngII, CTGF, FGF, IGF, PDGF, TGF-β
Anti-inflammatory	IL-10, TGF-β
Protective	BMP-7, HGF, VEGF
Chemokines	ENA78, IL-8, IP-10, MCP-1, MIF, MIP-1, osteopontin, RANTES
Cell surface receptors	CD40, CD48, MHC I & II
Complements	C2, C3, C3a, C4, C5a, factor H
Matrix	Biglycan, collagens, decorin, hyaluronan, fibronectin, laminin, TSP-1
Matrix modifiers	PAI-1, plasmin, MMP, SPARC, TIMP
Reactive oxygen species	

rich in cysteine (SPARC). Similarly, TEC near the urinary pole, injured directly or indirectly by the mediators released from the sublethal injured podocytes, may amplify the inflammatory cascade on podocytes. A vicious cycle is formed, which promotes an irreversible lethal response of the already sublethally injured podocytes. The potential mediators produced by the activated TEC are shown in table 1, which include pro-inflammatory (IFN-γ, IL-1β, IL-6, TNF-α) and pro-fibrotic (AngII, connective tissue growth factor (CTGF), fibroblast growth factor (FGF), insulin-like growth factor (IGF), PDGF, TGF-β) cytokines [5]. At the same time, TEC might also produce immunoregulatory molecules including IL-10, TGF-β, bone morphogenetic protein-7 (BMP-7), hepatocyte growth factor (HGF), or VEGF, that control the pro-inflammatory or pro-fibrotic cytokine response. These factors regulate the production of chemokines by TEC, which induce monocyte-macrophage and inflammatory cells recruitment in the interstitium. The chemokines released by TEC include neutrophil-activating protein 78 (ENA78), IL-8, IFN-γ-induced protein-10 (IP-10), MCP-1, macrophage inhibitory factor (MIF), MIP-2, osteopontin or regulated upon activation normal T-cell expressed and secreted (RANTES). Podocytes express chemokine receptors and produce ROS in response to the chemokine stimulation from TEC [14]. In turn, ROS stimulates GM-CSF and cell cycle checkpoint protein production by podocytes and eventually induces apoptosis.

Repair Mechanism

Kidney tubules can regenerate after sublethal or transient injury and TEC retain significant phenotypic plasticity in response to a variety of stimuli. However, the identity of tubular repairing cells has been debated. Although stem cells have been suggested to be the right candidate, later findings have suggested that neither the specialized progenitors from bone marrow, nor the telomerase expressing papillary epithelial cells, contribute to the repair of the injured kidney tubule [31]. This favors the traditional model of endogenous kidney repair by TEC that survive injury, and the process involves dedifferentiation of endogenous differentiated TEC and the activation of renal resident stem cells. Recent studies have discovered subsets of glomerular parietal epithelial cells (PEC), which may convert to podocytes and exhibit stem cell function [32]. It has been proposed that the PEC located at the urinary pole, are podocyte stem cells. After podocytic injury and cellular detachment, PEC may migrate towards the areas of injury along Bowman's capsule. PEC expressing CD24 and CD133 are also found to be located near the urinary pole, and when cultured in vitro, these CD24+CD133+ PEC exhibited high proliferative power and could differentiate into cells that expressed TEC markers. Further works need to be done to elucidate the regulation and exact mechanism of repair of podocytes or TEC injury by PEC.

Conclusion

A schematic diagram illustrates the possible crosstalk between podocytes and TEC is shown in figure 1. Severe injury to podocytes, together with the possible amplification by the secretory products from the resident glomerular and the infiltrating inflammatory cells, contributes to the apoptosis and detachment of the podocytes. This is followed by a series of detrimental events including proteinuria, podocytic cast formation, glomerular tuft adhesion and tubular obstruction. These events cause subsequent activation of TEC through excessive uptake of filtered macromolecules. At the same time, the filtered bioactive macromolecules in the ultrafiltrate exert pro-inflammatory and pro-apoptotic action on TEC. Together with the hypoxic or oxidative stress, and the recruitment of inflammatory cells to the tubulointerstitial region through chemokines secreted by the injured TEC, there is further amplification of the deleterious immune responses. As a consequence, TEC undergo apoptosis and EMT occurs, and finally tubular atrophy and fibrosis will develop. On the other hand, mild injury to podocytes may also impose stress on TEC near the urinary pole. The TEC will release circulating mediators in response to these stress and lead to progressive transition to the irreversible podocytic apoptosis and cellular detachment. PEC have the potential to repair the sublethal injured podocytes or TEC, and the relevant repair mechanisms and regulatory pathways are unknown at present and further investigation is warranted.

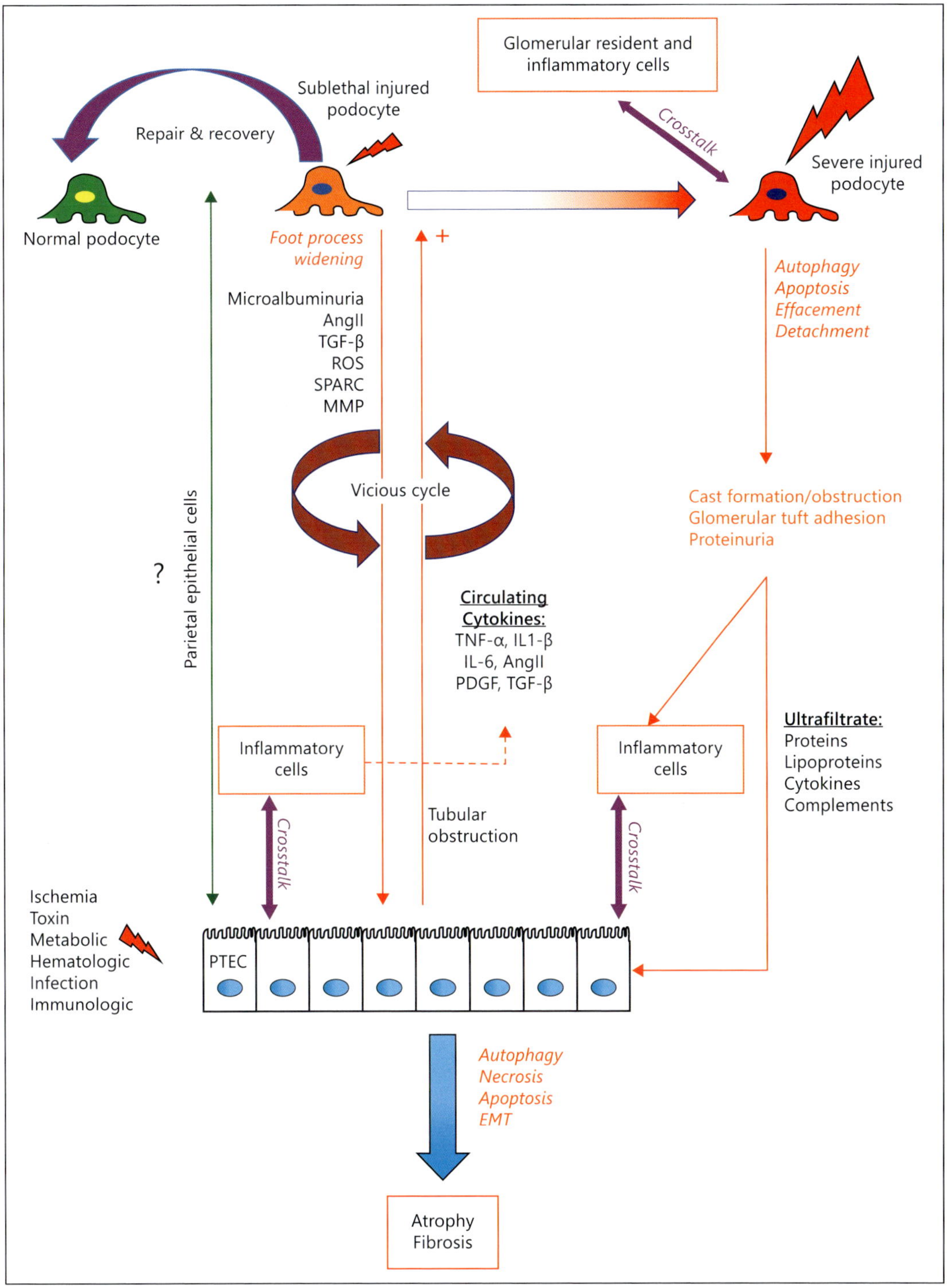

Fig. 1. Schematic representation of the crosstalk between podocytes and TEC.

Acknowledgments

The study was supported by the National Basic Research Program of China 973 program No. 2012CB517600 (No. 2012CB517606), a General Research Fund from the Research Grants Council of Hong Kong (grant No. HKU 768910M) and was partly supported by L&T Charitable Foundation and the House of INDOCAFE. We apologize to those investigators whose work was not cited due to space limitations.

Disclosure Statement

The authors have no conflicts of interest to disclose.

References

1 Garg P, Holzman LB: Podocytes: gaining a foothold. Exp Cell Res 2012;318:955–963.
2 Lai KN, et al: Podocyte injury induced by mesangial-derived cytokines in IgA nephropathy. Nephrol Dial Transplant 2009;24:62–72.
3 Liang XB, et al: Angiotensin type 1 receptor blocker restores podocyte potential to promote glomerular endothelial cell growth. J Am Soc Nephrol 2006;17:1886–1895.
4 Barisoni L, et al: Advances in the biology and genetics of the podocytopathies: implications for diagnosis and therapy. Arch Pathol Lab Med 2009;133:201–216.
5 Nangaku M: Mechanisms of tubulointerstitial injury in the kidney: final common pathways to end-stage renal failure. Intern Med 2004; 43:9–17.
6 Mundel P, Reiser J: Proteinuria: an enzymatic disease of the podocyte? Kidney Int 2010;77: 571–580.
7 Menzel S, Moeller MJ: Role of the podocyte in proteinuria. Pediatr Nephrol 2011;26: 1775–1780.
8 Thurman JM, Holers VM: The central role of the alternative complement pathway in human disease. J Immunol 2006;176:1305–1310.
9 Gaarkeuken H, et al: Complement activation by tubular cells is mediated by properdin binding. Am J Physiol Renal Physiol 2008; 295:F1397–F1403.
10 Rangan GK, et al: Differential effects of albumin on cytokine gene expression in proximal tubular epithelial cells. Nephrol Dial Transplant 2005;20:1013–1014.
11 Tang S, et al: Albumin stimulates interleukin-8 expression in proximal tubular epithelial cells in vitro and in vivo. J Clin Invest 2003;111:515–527.
12 Liu J, et al: Multiplex bead analysis of urinary cytokines of type 2 diabetic patients with normo- and microalbuminuria. J Immunoassay Immunochem 2010;31:279–289.
13 Abbate M, et al: How does proteinuria cause progressive renal damage? J Am Soc Nephrol 2006;17:2974–2984.
14 Shankland SJ: The podocyte's response to injury: role in proteinuria and glomerulosclerosis. Kidney Int 2006;69:2131–2147.
15 Lee HS, Song CY: Effects of TGF-β on podocyte growth and disease progression in proliferative podocytopathies. Kidney Blood Press Res 2010;33:24–29.
16 Gentle ME, et al: Epithelial cell TGF-β signaling induces acute tubular injury and interstitial inflammation. J Am Soc Nephrol 2013;24: 787–799.
17 Tipping PG: Are podocytes passive or provocative in proteinuric glomerular pathology? J Am Soc Nephrol 2008;19:651–653.
18 Li JH, et al: Role of TGF-β signaling in extracellular matrix production under high glucose conditions. Kidney Int 2003;63:2010–2019.
19 Burns WC, et al: Angiotensin II mediates epithelial-to-mesenchymal transformation in tubular cells by ANG 1–7/MAS-1-dependent pathways. Am J Physiol Renal Physiol 2010; 299:F585–F593.

20 Chevalier RL, Forbes MS: Generation and evolution of atubular glomeruli in the progression of renal disorders. J Am Soc Nephrol 2008;19:197–206.
21 Gluhovschi C, et al: What is the significance of HLA-DR antigen expression in the extraglomerular mesangium in glomerulonephritis? Hum Immunol 2012;73:1098–1101.
22 Pontrelli P, et al: CD40L proinflammatory and profibrotic effects on proximal tubular epithelial cells: role of NF-κB and lyn. J Am Soc Nephrol 2006;17:627–636.
23 Verzola D, et al: Oxidative stress mediates apoptotic changes induced by hyperglycemia in human tubular kidney cells. J Am Soc Nephrol 2004;15(suppl 1):S85–S87.
24 Forbes JM, et al: Oxidative stress as a major culprit in kidney disease in diabetes. Diabetes 2008;57:1446–1454.
25 Leung JC, et al: Oxidative damages in tubular epithelial cells in IgA nephropathy: role of crosstalk between angiotensin II and aldosterone. J Transl Med 2011;9:169.
26 Burton CJ, et al: Secretion of chemokines and cytokines by human tubular epithelial cells in response to proteins. Nephrol Dial Transplant 1999;14:2628–2633.
27 Ramesh G, Reeves WB: TNF-α mediates chemokine and cytokine expression and renal injury in cisplatin nephrotoxicity. J Clin Invest 2002;110:835–842.
28 Bonventre JV, Yang L: Cellular pathophysiology of ischemic acute kidney injury. J Clin Invest 2011;121:4210–4221.
29 Akcay A, et al: Mediators of inflammation in acute kidney injury. Mediators Inflamm 2009, p 137072.
30 Lee SJ, et al: Podocytes express IL-6 and lipocalin 2/neutrophil gelatinase-associated lipocalin in lipopolysaccharide-induced acute glomerular injury. Nephron Exp Nephrol 2012;121:e86–e96.
31 Humphreys BD, et al: Repair of injured proximal tubule does not involve specialized progenitors. Proc Natl Acad Sci USA 2011;108: 9226–9231.
32 Shankland SJ, et al: Glomerular parietal epithelial cells in kidney physiology, pathology, and repair. Curr Opin Nephrol Hypertens 2013;22:302–309.

Prof. Sydney C.W. Tang
Department of Medicine, University of Hong Kong
Room 415, 4th Floor, Professorial Block, Queen Mary Hospital
Hong Kong, SAR (China)
E-Mail scwtang@hku.hk

transported across the IMM by their carrier proteins. The IMM contains a large number of proteins required for complicated biochemical reactions, including specific carrier proteins for metabolites and intermediate products (e.g. phosphoric acid, glutamine, ornithine, nucleotides, and ions), translocase of the inner membrane, enzymes involved in electron transfer and ATP synthesis, and proteins controlling the fission and fusion of mitochondria. The IMM folds inward to form cristae, thereby greatly expanding the surface area of the IMM. The cristae are studded with stalked, small, round spheres, known as elementary particles. These particles contain ATP synthase, which uses energy from the ETC to synthesize ATP [5].

The OMM and IMM form two compartments: the intermembrane space and the mitochondrial matrix. The intermembrane space, the space between the OMM and IMM, is full of amorphous liquid. As a result of the high permeability of the OMM and poor permeability of the IMM, the content of the intermembrane space is similar to that of the cytoplasm, containing various biochemical substrates, soluble enzymes, and cofactors. The mitochondrial matrix, the space enclosed by the IMM, contains mtDNA, RNA, and a large number of proteins, such as the enzymes involved in biochemical reactions (e.g. Krebs cycle, fatty acid oxidation, and amino acid degradation) [2].

Recent electron microscopy studies have suggested a three-dimensional model of mitochondria, featuring the following characteristics [4]. First, the OMM is structurally and functionally linked to endoplasmic reticulum (ER) membranes, cytoskeleton, and other cellular components, thus forming a network. Second, there are random, distributed contact sites between the OMM and IMM. Third, the cristae are not formed by simple invagination of the IMM. Instead, to form the cristae, the peripheral portion of the IMM (known as the inner boundary membrane) is connected to the cristae membranes via crista junctions. Fourth, the inner cristae membranes are tubular or flat sac-like structures that can be interconnected or fused. This new three-dimensional model is critical for studies on molecular movement in mitochondria, mitochondrion-ER membrane connections, and the formation of mitochondrial networks in cells.

Mitochondrial DNA

Human mitochondria contain their own DNA (mtDNA) that encodes 37 genes, all of which are essential for normal mitochondrial functions (fig. 1). Thirteen of these genes encode enzymes involved in OXPHOS, and the remaining genes encode transfer RNA (tRNA) and ribosomal RNA.

Most human cells contain hundreds to thousands of mtDNA copies per cell. In comparison, the nucleus contains only two copies of nuclear DNA. Consequently, mtDNA and nuclear DNA are remarkably different in their genetic pat-

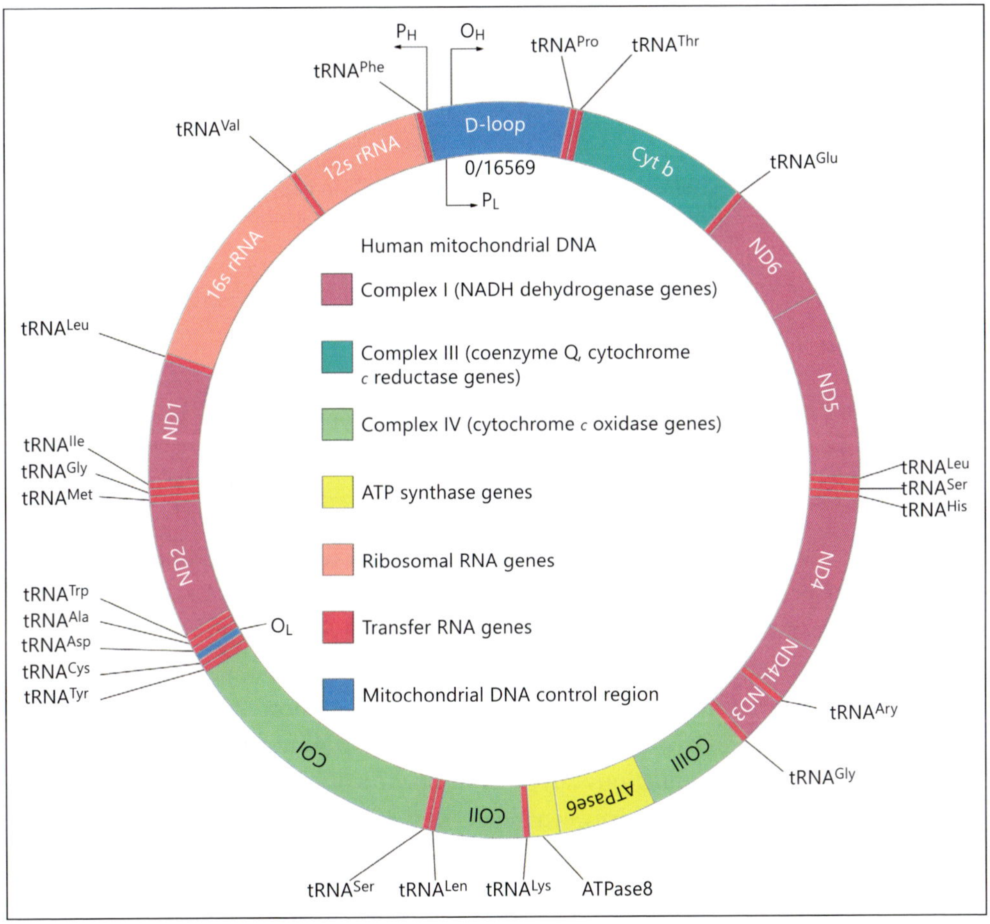

Fig. 1. Gene map of human mtDNA. The 16,569 bp human mtDNA comprises 13 mRNA, 2 rRNA (12S and 16S), and 22 tRNA coding genes. mRNA genes are shown with the codes of the corresponding ETC complexes I, III, IV, and V. mtDNA contains no introns. The only non-coding regions are the replication origin of the light strand (O_L) and a structure called the displacement loop (D-loop) that contains the replication origin of the heavy strand (O_H) and the heavy and light strand promoters (P_H and P_L, respectively).

terns (table 1). The genetics of mtDNA deviates from the traditional mendelian inheritance in four aspects. First, mtDNA follows maternal inheritance. Because a spermatozoon contains only a negligible amount of cytoplasm, essentially all mtDNA in the offspring is derived from the ovum. As a result, mtDNA shows a maternal inheritance without DNA recombination. Thus, individuals with identical mtDNA sequences are undoubtedly offspring of a common female ancestor. Nevertheless, recent studies using polymerase chain reaction found that the spermatozoon also contributes mtDNA to the zygote. Second, mtDNA inheri-

Table 1. Differences between mtDNA and nDNA

DNA characteristics	mtDNA	nDNA
Size	16,568 bp	3×10^9 bp
Shape	double-stranded loop	linear double strand
Number per cell	hundreds to thousands	23 in haploid cells, 46 in diploid cells
DNA-binding proteins	none	histone/non-histone
Encoding DNA, %	93	3
Repeating DNA, %	low	high
Transcription	continuous transcription of multiple genes	individual transcription
Introns	absent	present
Codons	start codon: AUA stop codons: AGA, AGG	start codon: AUG stop codons: TAA, TGA, TAG

tance displays heteroplasmy. Because hundreds to thousands of mtDNA copies are present in a cell, mutant mtDNA and wild-type mtDNA exist in the cell at disparate proportions [6]. A normal individual carries only a small fraction of mutant mtDNA (0.2–2%) [7]. The mtDNA from the spermatozoon contributes to this heteroplasmy [8]. Third, mtDNA mutation shows a threshold effect. Mutations of mtDNA lead to reductions in OXPHOS activity and ATP synthesis. Whether such mutations can generate phenotypic effects in a given tissue is determined by the ratio between mutant and normal mtDNA, and the dependence of the tissue on ATP synthesized by mitochondria. Mutations may result in functional changes of tissues or organs only after reaching a threshold proportion. Finally, during cell replication, mutant and wild-type mtDNA are allocated to daughter cells stochastically. As a result, heterozygotes can form both wild-type and mutant homozygotes. However, as the replication of mutants usually predominates over the wild-type counterparts, mutations are likely to accumulate.

The mtDNA is exposed to ROS resulting from OXPHOS without the protection of histone. Moreover, it consists almost entirely of coding regions lacking repair mechanisms. As a result, it is susceptible to damage and mutations [9–12]. The main causes of mtDNA damage include ROS, UV, ionizing irradiation, alkylating agents, base analogs, modifier-induced base-pair variations, and aging. Even without external stimulants, mtDNA also naturally undergoes damage, such as base mismatches during replication and spontaneous base changes. The types of damage include single-strand breakage, double-strand breakage, and inter-strand cross-linking. Despite this high probability of damage, mtDNA is equipped with only inadequate and inefficient repair mechanisms. The encoded products are important components of OXPHOS enzyme complexes, and their expression levels are determined by the number of mtDNA copies. Consequently, mtDNA mutations may cause mitochondrial dysfunction, including reduced ATP synthesis, elevated intracellular calcium levels (resulting from calcium pump inactivation), activated phospholipases, and decomposition of membrane phospholipids [13, 14].

Functions

OXPHOS and ATP Synthesis

Known as the 'powerhouses' of cells, mitochondria are organelles where nutrients (i.e. saccharides, lipids, and amino acids) are oxidized to generate energy. The primary function of mitochondria is to perform OXPHOS and, thereby, synthesize ATP. Approximately 90% of the energy required by a cell is generated in the mitochondria by OXPHOS. Nutrients such as saccharides and lipids are glycolyzed in the cytoplasm into pyruvic acid and fatty acid. These acids enter mitochondrial elementary particles, form acetyl coenzyme A via a series of catabolic reactions, and subsequently enter the Krebs cycle. Energy is released as a result of the interactions between the Krebs cycle and the ETC. The Krebs cycle produces primarily protons and electrons (NADH and FADH), as well as ATP. The protons and electrons then enter the ETC, producing more ATP. Therefore, ATP synthesis requires two coupled processes: electron transport and proton transport. OXPHOS includes two components: ETC and ATP synthase.

Located on the IMM, the ETC comprises five protein complexes: complex I (NADH dehydrogenase), complex II (succinate dehydrogenase), complex III (ubiquinone-cytochrome *c* oxidoreductase), complex IV (cytochrome *c* oxidase), and complex V (ATP synthase) (fig. 2). Additionally, the ETC requires two small-molecule electron carriers: ubiquinone and cytochrome *c*. Complexes I, III, and IV are proton pumps, responsible for the pumping of protons (produced by the Krebs cycle) from the mitochondrial matrix to the intermembrane space. Because the IMM is highly impermeable to ions, the proton concentration in the intermembrane space is higher than that in the mitochondrial matrix, consequently forming a pH gradient and an electrical potential gradient across the IMM (thus a transmembrane electrochemical gradient). Along this electrochemical gradient, protons flow through complex V (ATP synthase) on the IMM to return to the mitochondrial matrix. This reflux changes the conformation of complex V and drives the synthesis of ATP from ADP and inorganic phosphate. As a result, the energy extracted from the oxidation of hydrogen carriers is stored in the form of ATP.

ROS Generation

In an organism, more than 90% of the oxygen used is consumed in the mitochondria, including 1–2% used for ROS generation. Thus, although an essential nutrient, oxygen has two very different actions. First, as the ultimate electron

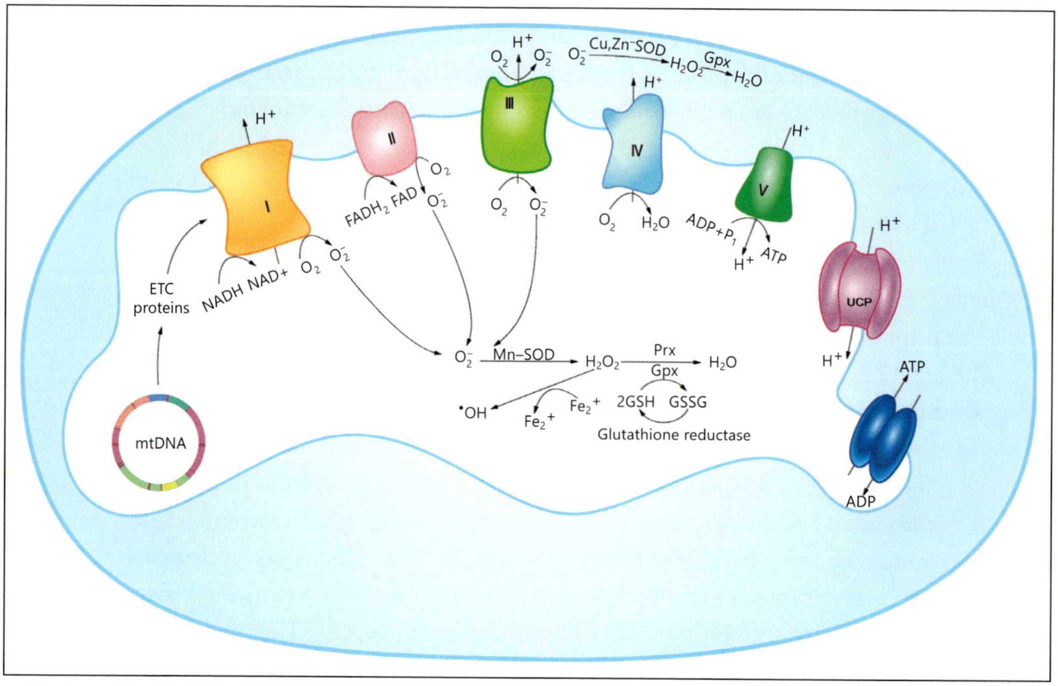

Fig. 2. OXPHOS, superoxide production, and scavenging pathways in normal mitochondria. The energy for ATP synthesis is generated by the oxidation of NADH and FADH2 through the ETC (ETC). The ETC is a sequence of electron transport carriers (complexes I, II, III, and IV), which transport electrons gradually to O_2. Superoxide (O_2^-) generated by the ETC is mostly released to the matrix at complex I, and both the matrix and intermembrane space at complex III. Superoxide is dismutated to H_2O_2 by Cu, Zn-SOD in the intermembrane space, and by Mn-SOD in the matrix. H_2O_2 can react with Fe^{2+} to generate the highly reactive hydroxyl radical (•OH) via Fenton chemistry. H_2O_2 is reduced to H_2O by peroxiredoxin (Prx) or glutathione peroxidase (Gpx) using GSH, and the resultant oxidized glutathione (GSSG) is reduced back to GSH by glutathione reductase. P_i stands for inorganic phosphate.

acceptor in the ETC, oxygen participates in energy metabolism essential for life. On the other hand, oxygen can be converted by a variety of reactions into radicals, which may lead to cell injuries, human disease, and ageing. Studies have shown that mitochondria are major sites of intracellular ROS generation. A mitochondrion can produce 3×10^7 superoxide radicals daily, accounting for 95% of the ROS in vivo. As shown in figure 2, electrons 'leaking' from the ETC combine with oxygen molecules (i.e. single-electron reduction) to form superoxide anions (O_2^-), a precursor of ROS. Mitochondrial O_2^-, formed primarily in complexes I and III, can be catalyzed by superoxide dismutase (SOD) to disproportionate into hydrogen peroxide (H_2O_2). Subsequently, H_2O_2 can form hydroxyl radicals ($^{\cdot}OH$) via the Fenton reaction. $O_2^{\bar{}}$, H_2O_2, and $^{\cdot}OH$ are collectively known as ROS. Mitochondrial $O_2^{\bar{}}$ can also combine with protons to form uncharged $HOO^{\cdot}$ radicals, which subsequently enter mitochondrial membranes and react with unsaturated fatty acid components of membrane lipids to yield lipid radicals. Mitochondria also produce nitric oxide (NO), which can combine with $O_2^{\bar{}}$ to form peroxynitrite ($ONOO^-$), a stable anionic radical. The conjugate acid of this radical, peroxynitric acid, can be transported across membranes over a long distance.

The presence of a low concentration of ROS is physiologically normal. In fact, ROS function as important second messengers. An excessive generation of ROS is, however, injurious to mtDNA, potentially leading to impaired ETC functions, reduced ATP synthesis, mitochondrial dysfunction, and cell injuries, and even apoptosis (fig. 3). ROS play key roles in the initiation and modulation of cell apoptosis.

A mitochondrion is equipped with antioxidant mechanisms to control the concentrations of mitochondrial ROS (fig. 2). First, the mitochondrion scavenges $O_2^{\bar{}}$ with superoxide dismutase 2 (SOD_2) (manganese SOD, Mn-SOD). Second, the mitochondrion can inhibit single-electron reduction of O_2 by decoupling. During the switch of the mitochondrial respiration state from III to IV, ADP is depleted and ATP accumulates. Consequently, the main pathway of O_2 consumption is blocked, resulting in an elevated O_2 concentration. Meanwhile, the main targets of O_2 single-electron reduction (e.g. flavoprotein, ubiquinone, cytochrome *b*) are in the reduced state. This hyperoxic environment and reduced state of the ETC drive the leakage of electrons to single-electron reduction of O_2, generating $O_2^{\bar{}}$. In response, the mitochondrion can decouple this side-path by increasing the proton reflux across the IMM, thereby stimulating O_2 consumption and reducing $O_2^{\bar{}}$ generation. Third, when a mild decoupling is insufficient, the accumulation of ROS triggers the mitochondrion to open mitochondrial permeability transition pores (MPTP) in its IMM. With large pore sizes, MPTP enable protons to reach equilibrium across the IMM, and also allows equilibrium of ETC substrates between the cytoplasm and mitochondrial matrix. With the reduction of the transmembrane proton gradient, the respiratory rate of the ETC reaches the maximum, driving a rapid consumption of oxygen and deceleration of ROS generation. With the resolution of ROS accumulation, the MPTP are then closed. A sustained excessive accumulation of ROS may induce a prolonged opening of MPTP, potentially triggering apoptosis. Finally, cytochrome *c* also plays critical roles in the mitochondrial antioxidant system. The oxidized form of cytochrome *c* can directly oxidize $O_2^{\bar{}}$ to water. Additionally, cytochrome *c* partially scavenges $O_2^{\bar{}}$ and H_2O_2. For rapid oxidation of $O_2^{\bar{}}$ in a mitochondrion, cytochrome *c* should

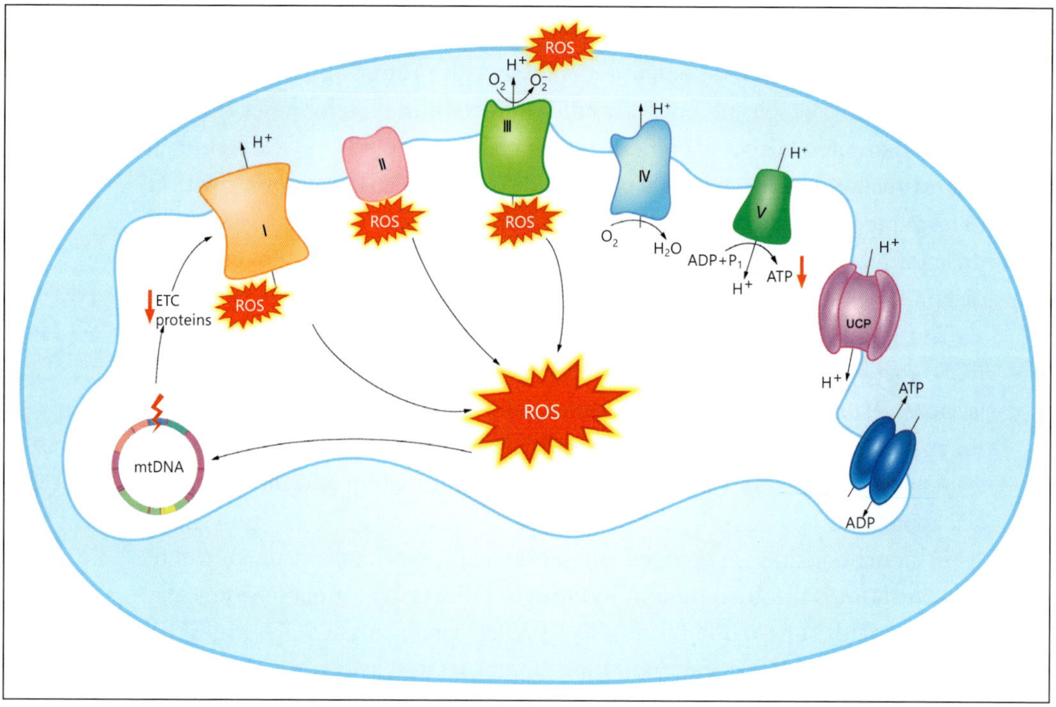

Fig. 3. OXPHOS, superoxide production, and scavenging pathways in dysfunctional mitochondria. In dysfunctional mitochondria, the electron transport carriers (complexes I, II, III, and IV) do not efficiently transfer electrons, leading to increased ROS production and decreased ATP production. Excessive ROS can attack all molecules, including mtDNA, and consequently causes a decrease in mtDNA and altered expression of mitochondrial proteins essential for ETC and ATP synthesis. Defects in mitochondrial proteins affect ETC activity culminating in a vicious cycle of ROS production.

be released into the intermembrane space in a soluble form. When MPTP remain open, cytochrome *c* enters the cytoplasm to activate the apoptotic signaling pathway.

Calcium Homeostasis

Intracellular calcium (Ca^{2+}) homeostasis plays a fundamental role in cell metabolism, proliferation, differentiation and cell death. Mitochondria act as a spatial Ca^{2+} buffer, which regulates the local Ca^{2+} concentration in many cells [15]. Ca^{2+} uptake is via a ruthenium red-sensitive mitochondrial Ca^{2+} uniporter (MCU) and is driven by the membrane potential [16]. Ca^{2+} accumulation by MCU does not proceed until electrochemical equilibrium is reached due to the activity of two main efflux pathways, the Na^+-dependent and the Na^+-independent mechanisms [17]. The existence of the mitochondrial Ca^{2+} transport molecular machinery supports the general consensus that mitochondria can actively and rapidly change their Ca^{2+} concentrations and participate in cellular homeostasis [18].

Ca^{2+} is a key regulator of mitochondrial function and acts on several levels within the organelle. Three crucial metabolic enzymes within the matrix, pyruvate, α-ketoglutarate and isocitrate dehydrogenase, are activated by elevated mitochondrial Ca^{2+}, resulting in faster respiratory chain activity and higher ATP output. Thus, mitochondrial ATP output can be changed to meet cellular ATP demands [19]. In contrast to the beneficial effects of Ca^{2+}, mitochondrial Ca^{2+} is also involved in the regulation of cell death pathways. Ca^{2+} overload, in conjunction with the accumulation of ROS, favors the sustained opening of MPTP, which causes the rapid collapse of the membrane potential and the swelling of mitochondria, leading to the release of cytochrome *c* [16]. Recently, mitochondrial Ca^{2+} signals have been shown to play a major role in the regulation of autophagy. The absence of mitochondrial Ca^{2+} transfer results in the activation of AMP-activated protein kinase, which in turn activates mammalian target of rapamycin (mTOR)-independent prosurvival autophagy [20].

Induction of Cell Apoptosis

When cells experience injurious stimuli and antioxidant mechanisms (e.g. decoupling, transient MPTP opening, SOD_2), or fail to prevent ROS accumulation, the risk of damage to mtDNA (and even nuclear DNA) increases substantially. The damage is able to accumulate and can be inheritable. Excessive accumulation of damaged cells impairs the normal functions of tissues and organs. Therefore, at certain threshold levels, the organism initiates apoptotic mechanisms to eliminate severely damaged cells. Studies have discovered three pathways for inducing cell apoptosis: death receptor pathway, mitochondrial pathway, and ER membrane pathway [10, 21]. In mammalian cells, the mitochondrial pathway is the most common, with mitochondria being the key to this pathway. When ROS concentrations reach high levels, cytochrome *c* dissociates from the IMM into the intermembrane space for defense. In case of sustained ROS elevation, MPTP are opened and the transmembrane electrical potential gradient decreases, initiating early apoptosis. With the opening of MPTP (with large pores), water and other small molecules migrate from the cytoplasm to the mitochondrial matrix, thereby causing matrix dilution, straightening of cristae, and the rupture of the OMM. Cytochrome *c* is also released into the cytoplasm and binds to apoptosis activation factor 1 and procaspase-9, forming apoptosomes. Caspase-9 is activated, which activates caspase-3, and then the caspase protease system. Finally, a series of hydrolases are triggered, causing cell death. These events form the classic pathway of mitochondria-mediated cell apoptosis.

The release of mitochondrial cytochrome *c* into the cytoplasm is common to various paths of apoptosis. This release is controlled by Bcl-2, a family of proteins residing on mitochondrial outer membranes. Members of this family (e.g. Bcl-2, Bcl-W, Bcl-X_L) inhibit apoptosis, whereas other members (e.g. Bax, Bad, Bid) activate this suicidal process.

Mitochondrial Dysfunction and Podocyte Injury

The main pathophysiological changes associated with mitochondrial dysfunction include increased ROS generation, impaired ATP synthesis, elevated cytoplasmic Ca^{2+} concentrations, cellular energy metabolism dysfunction, activation of proteases and endonucleases, and morphological changes of mitochondria. Ultimately, these changes lead to cell apoptosis.

Podocytes are highly specialized epithelial cells lining the urinary surface of the glomerular capillary tuft. Podocyte injury is considered a key factor in the development of many forms of human and experimental glomerular diseases, such as diabetic nephropathy, minimal change disease, focal segmental glomerulosclerosis (FSGS), membranous glomerulopathy, and lupus nephritis [22]. The podocyte consists of a cell body, major processes, secondary processes and finely interdigitating foot processes. In addition, microtubules and intermediate filaments are present in the cell body and major processes, and actin is the major structural protein in foot processes. Both actin and microtubules are indispensable for the maintenance of the architecture of podocytes [23]. Large amounts of ATP as the energy supply may be required for all these processes, which suggest an important role of mitochondria, the major site of ATP synthesis in eukaryotic cells, in maintaining podocyte energy homeostasis. Several studies have shown that mitochondrial dysfunction contributes to podocyte injury [24, 25]. Another study demonstrated that mitochondrial dysfunction was an early event in podocyte injury [26].

mtDNA and Podocytes

Since the first two pathogenic mutations in mtDNA were reported in 1988, there has been an explosion in the recognition of distinct nuclear and mitochondrial genetic causes of mitochondrial disease [27]. Over the past two decades, a vast amount of mitochondrial-associated gene abnormalities have been reported to be involved in several renal diseases, including tubular disorders, chronic tubulointerstitial nephritis, cystic renal disease, and glomerular diseases [28].

It is now well known that FSGS is a podocyte disease. Several studies have demonstrated FSGS and proteinuria in patients with an A-to-G transition at mtDNA position 3243 in the tRNALeu(UUR) gene [29]. Some cases with different mtDNA deletions have been reported also [30]. Among them, 1 patient presented with nephrotic syndrome at 4 months of age. Podocytes from renal biopsy showed proliferation of pleomorphic mitochondria [31]. During the development of glomerulopathy in a puromycin aminonucleoside nephrosis (PAN) mouse model, the copy number of a 4,834 bp deletion (del-mtDNA) increased in FSGS phase, and the mtDNA encoding the cytochrome *c* oxidase subunit I protein level reduced, suggesting mitochondrial dysfunction by mtDNA depletion plays a key role in the pathogenesis of FSGS in PAN [25]. Besides mtDNA disorders, nuclear-encoded mitochondrial diseases are associated with renal phenotypes [32]. Coenzyme Q10 (CoQ10, ubiquinone) is an essential component of the mitochondrial ETC, and is a lipid-soluble antioxidant molecule. The histological change in CoQ10-related diseases with glomerular involvement varies from FSGS to collapsing glomerulopathy. Electron microscopy generally shows numerous dysmorphic mitochondria in the cytoplasm of podocytes. Decaprenyl-diphosphate synthase subunit 2 (PDSS2) is a part of the CoQ10 biosynthesis pathway. Podocyte-specific PDSS2 gene knockout mice resulted in proteinuria and podocyte foot process effacement [33]. Furthermore, Weiher et al. [34] reported mice carrying MPV17 null alleles developed FSGS-like glomerular pathology and proteinuria. The MPV17 gene encodes a mitochondrial inner membrane protein that is implicated in the preservation of mitochondrial homeostasis by controlling mtDNA maintenance and OXPHOS activity. Suppression of the function of MPV17/HTRA2 complexes in podocytes by TGF-β might induce mitochondrial dysfunction and apoptosis associated with mtDNA depletion [35].

ATP Synthesis in Podocytes

Mitochondria are responsible for creating more than 90% of the energy needed by the body to sustain life and support growth. Cell injury and even cell death happens when mitochondria fail to generate enough energy [36]. The kidney consumes a large amount of energy, supplied as ATP, so any disruption in this supply is likely to cause renal dysfunction [37].

Podocytes are highly specialized and terminally differentiated cells. They are likely to have high energy consumption to maintain their appropriate structure and that of their foot processes [38]. During intracellular energy depletion by glucose starvation in podocytes, induction of ER stress was demonstrated by increased expression of the stress-inducible chaperone GRP78, and the inhibition of nephrin biogenesis was confirmed by expression of immature 155 kDa nephrin [39]. It was also reported that dexamethasone rescued this interfered trafficking of nephrin under ER stress through binding to its receptor and stim-

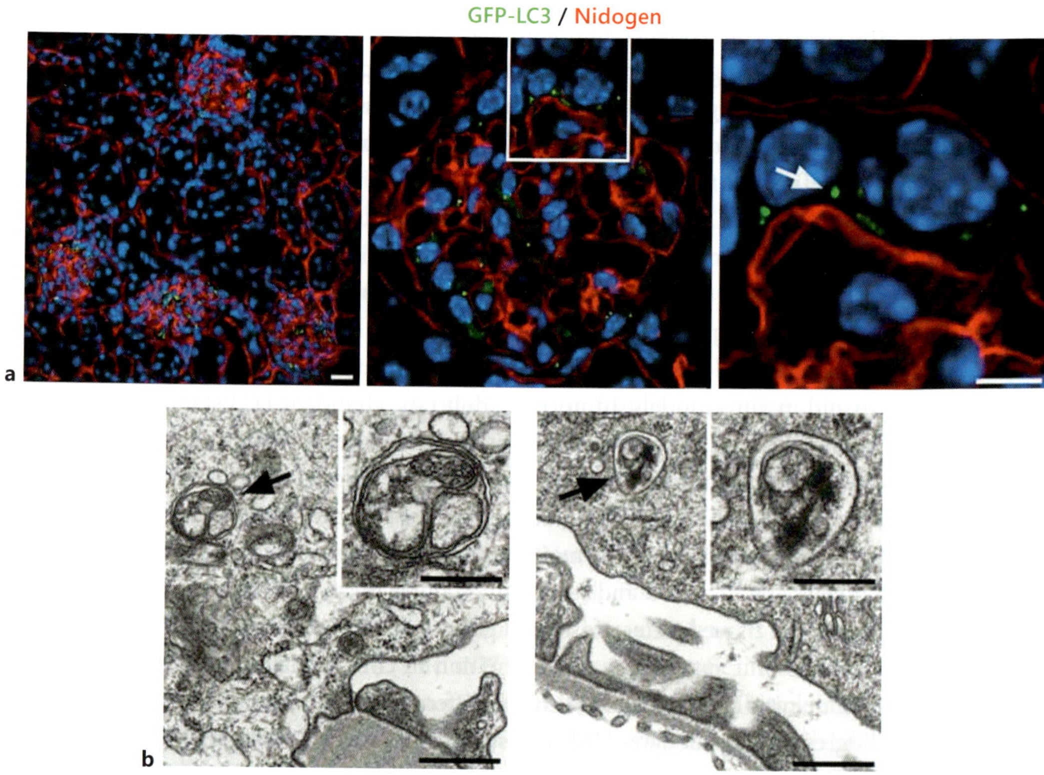

Fig. 2. Glomerular podocytes display high levels of autophagy under basal conditions. **a** In vivo analysis of autophagy using GFP-LC3 transgenic mice (arrow indicates autophagosome (green dots) within the podocyte; nidogen (red) is a marker for the basement membrane). **b** Electron microscopy showing autophagosome (arrow, left) and autolysosome (arrow, right) in podocytes of mice. The figure is adopted from Hartleben et al. [18], with the permission from American Society for Clinical Investigation.

mic LC3-I to membrane-bound LC3-II conversion [19]; thus, the *Atg5*/*Atg7*-independent alternative pathway might compensate for the loss of Atg5-mediated autophagy in either the conventional or the conditional podocyte-specific *Atg5* knockout mice [18].

Basal autophagy is important for cellular homeostasis and quality control through the constitutive turnover of cytoplasmic components including long-lived proteins and organelles. This occurs in both cells with remarkably high capacity of regeneration such as hepatocytes [20] and cells with limited capacity of regeneration such as neurons [21]. Podocytes are unique in that they are terminally differentiated and relatively long-lived cells with high levels of basal autophagic activities. In a doxycycline-inducible podocyte-specific *Atg5* knockout model, acute induction of *Atg5* deletion in 12-week-old mice triggers a rapid onset of albuminuria. Nonetheless, up to 2–4 months after birth, constitutive podocyte-specific *Atg5* knockout mice are indistinguishable from wild-type littermates, as indicated by renal histology, glomerular ultrastructure, and urinalysis. These interesting observations suggest that in addition to the possibility of Atg5/Atg7-independent alternative autophagic pathway [19], other adaptive or defensive pathways such as the ubiquitin-proteasome pathway also might compensate for the constitutive loss of Atg5 [18]. Indeed, total proteasome activity is significantly increased in glomerular lysates of 8-month-old podocyte-specific *Atg5* knockout mice, consequently preventing an accumulation of polyubiquitinated proteins. Notably, inhibition of proteasome activity by bortezomib results in a drastic increase of albuminuria in 6-month-old podocyte-specific *Atg5* knockout mice. Moreover, inhibition of proteasome by MG132 in differentiated podocytes in vitro also significantly activates the conventional autophagic pathway as indicated by increased accumulation of LC3-II and formation of GFP-LC3 puncta [18]. Together, these results suggest that the basal autophagy in podocytes may act in concert with the proteasome pathway to maintain the structural and functional integrity of podocytes.

The role of basal autophagy in podocytes was further determined in a more recent study by Fang et al. [22]. Using paracellular permeability influx assay, they examined the filtration barrier function of podocytes and found that inhibition of autophagy by 3-MA in cultured mouse podocytes leads to an increased amount of albumin influx across the podocyte monolayer (leakage of albumin). Consistent with the functional results, the expression of podocyte slit diaphragm proteins such as nephrin, CD2AP, and podocin are also suppressed by 3-MA in both dose- and time-dependent manners. Similar observations were obtained by inhibition of basal autophagy with *Beclin1* siRNA [22]. Thus, this study further suggests that the high level of basal autophagy in podocytes plays an essential role in maintaining the architecture and function of the podocyte.

Autophagy in Podocyte Aging

Aging in humans and rodents progressively impairs the kidney, leading to renal dysfunction and increased vulnerability to stress. The age-related structural changes in the kidney, particularly in the glomerulus, are characterized by a loss of podocytes and glomerulosclerosis. However, the exact mechanisms underlying age-associated podocyte injury are unknown [23]. The accumulation of damaged proteins, lipids, and organelles is usually thought to account for the dysfunction of aging or aged cells. Studies have suggested that the accumulation of toxic metabolic wastes in the cell is caused, at least in part, by a decrease of protein degradation [24]. Autophagy defect with age observed in many types of cell and tissue has thus been proposed to contribute to different aspects of the phenotype of aging and to the pathogenesis of some age-related diseases such as neurodegenerative diseases, cancer, and metabolic disorders [25].

A recent study showed defective autophagy in aged kidneys in rats. The expression of LC3 and Atg7 is significantly decreased in old rats compared with young rats. As a result of autophagy inhibition, large amounts of p62 and polyubiquitin aggregates accumulate in aged kidneys, accompanied by an accumulation of damaged mitochondria and induction of oxidative stress. All of these alterations are typical features of aging cells [26]. Although direct evidence that aging podocytes have a decreased autophagy is still lacking, the role of autophagy in the development of age-related glomerulopathy has been examined in 20- to 24-month-old podocyte-specific *Atg5* knockout mice [18]. *Atg5* deletion in podocytes accelerated podocyte aging, as indicated by the accumulation of lipofuscin, protein aggregates, damaged mitochondria, and oxidized proteins. The result suggests that autophagy may be a direct determinant of cellular aging for long-lived podocytes. The excessive accumulation of intracellular waste induces ER stress and oxidative stress in the knockout mice. Under this condition, the aforementioned compensatory proteasome pathway that may clear protein aggregates in young *Atg5* knockout mice is significantly reduced in the old mice. Ultimately, the *Atg5*-deficient mice develop an injury and loss of podocytes and progressive glomerulosclerosis, as indicated by proteinuria, renal histology and podocyte ultrastructure [18]. These results highlight the fundamental role of autophagy for the long-time maintenance of glomerular podocytes. Based on these findings, the authors speculated that a gradual decrease of glomerular autophagic activity with age could play a major role in the decline of renal function in human aging kidneys [18]. It would be interesting to test this hypothesis. Signaling pathways leading to age-associated autophagy defect also need to be elucidated.

Role of Autophagy in Podocyte Disease

Primary Glomerular Diseases

Using in vivo and in vitro models of puromycin aminonucleoside (PAN) nephrosis (an experimental model of human nephrotic syndrome), an earlier study demonstrated the first evidence of autophagy in glomerular disease. Autophagy induction in glomeruli and in podocytes is correlated with the recovery from PAN nephrosis, suggesting a protective role of autophagy in preventing the development of and in recovering from nephrotic syndrome [16]. A number of studies have examined autophagy in podocytes in human renal biopsy samples from the patients with renal diseases such as pediatric IgA nephropathy [27–29]. Upregulation of autophagy in podocytes was further shown in *GFP-LC3* transgenic mice with albumin overload-induced proteinuria and in human renal biopsy specimens from patients with acquired proteinuric diseases including focal segmental glomerulosclerosis, membranous glomerulonephritis, and minimal change disease [18]. These observations are consistent with the findings that podocyte-specific *Atg5* knockout significantly sensitizes mice to glomerulosclerosis induced by either PAN or adriamycin. Moreover, BSA overload induces significantly higher transient albuminuria in podocyte-specific *Atg5* knockout mice compared with their wild-type littermates [18]. These data suggest that autophagy defends the integrity of podocytes against the development and progression of glomerular disease. In addition, upregulation of autophagy may also protect podocytes against passive Heymann nephritis (an animal model of membranous nephropathy) through repairing the cytoskeleton damage induced by ER stress [30].

Diabetic Nephropathy

Diabetic nephropathy is a type of proteinuric disease pathologically characterized by podocyte injury, GBM thickening, and mesangial expansion. The data on autophagy in diabetic nephropathy are limited and controversial.

Fang et al. [22] found that autophagy in podocytes is suppressed in streptozotocin-treated mice, in renal biopsy samples from diabetic patients, and in cultured mouse podocytes incubated with high glucose for 48 h. The defective autophagy induced by sustained high glucose is accompanied by podocyte injury, as indicated by reduced podocin expression and increased leakage of albumin. Rapamycin restores autophagy in podocytes and thereby rescues podocytes from injury. Furthermore, ER stress is induced by high glucose. Prolonged high glucose treatment (for 36–60 h) leads to a brief increase and then a decrease of eIF2α phosphorylation in podocytes, suggesting a decline of the adaptive unfolded protein responses. In contrast, the expression of the proapoptotic CHOP is maintained at an elevated level, indicating a switch from a cytoprotective to a cytotoxic ER stress response. Salubrinal or taurine-conjugated ursodeoxycholic acid partially reverses the switch, leading to restoration of autophagy and attenuation of podocyte injury, proteinuria, and glomerulopathy [22]. Together, these results highlight the renoprotective role of autophagy in podocytes in experimental models of diabetic nephropathy. Under these conditions, the induction of autophagy as a cytoprotective mechanism in response to ER stress may play an important role in the survival of the podocytes.

In another study, Ma et al. [31] mainly examined the short-term effects of high glucose on autophagy in cultured mouse podocytes. They found that high glucose incubation for 24 h induces autophagy in these cells, as indicated by increased LC3-II accumulation and autophagosome formation. High glucose-induced autophagy can be further enhanced by rapamycin but suppressed by 3-MA. Along with autophagy activation, oxidative stress is also induced, as indicated by increased generation of ROS and expression of MnSOD and catalase. Importantly, high glucose-induced autophagy is inhibited by the antioxidant N-acetylcysteine, supporting a role for oxidative stress in activating autophagy.

Apparently, these two studies demonstrate a time-dependent effect of high glucose on autophagy in podocytes. The underlying mechanism needs to be determined. Further studies using a mouse model of podocyte-specific *Atg* gene knockout would also be important to define the pathological role of autophagy in diabetic nephropathy.

Lysosomal Storage Diseases

Lysosomal storage diseases (LSDs) are a family of disorders caused by inherited gene mutations that perturb lysosomal homeostasis. Deficiencies in lysosomal enzymes or non-enzymatic lysosomal proteins lead to abnormal storage of macromolecules and lysosomal dysfunction. Since lysosomes play a central role in the quality control and selective clearance of intracellular components through multiple routes including autophagosomal-lysosomal pathway and endosomal-lysosomal pathway, both endocytosis and autophagy are usually affected in LSDs [32].

The mammalian homolog of yeast vacuolar protein sorting defective 34 (mVps34) is also known as class III phosphatidylinositol 3-kinase (PI3KIII). mVps34 specifically produces phosphatidylinositol-3-phosphate (PtdIns3P) to initiate autophagosome formation and also to regulate endocytosis [33]. We [34, 35] and others [34, 35] have recently demonstrated an essential role of mVps34 in regulating the autophagic and endocytic-lysosomal pathways and its significance in maintaining podocyte homeostasis. Podocyte-specific knockout of *mVps34* in mice causes striking podocyte vacuolization by 15 days of age before the onset of proteinuria; it induces marked foot process effacement with irregular and thickened GBMs and massive proteinuria by 3 weeks of age. These knockout mice develop severe glomerulosclerosis and interstitial fibrosis with significant growth retardation by 5–6 weeks of age and died within 9 weeks of age [34, 35]. Using immunoblotting of isolated renal glomerular lysates, we quantitatively determined that the membrane-bound LC3-II and even the cytosolic LC3-I were both increased; however, the ratios of LC3-II/LC3-I were elevated in podocyte-specific mVps34 knockout mice, suggesting

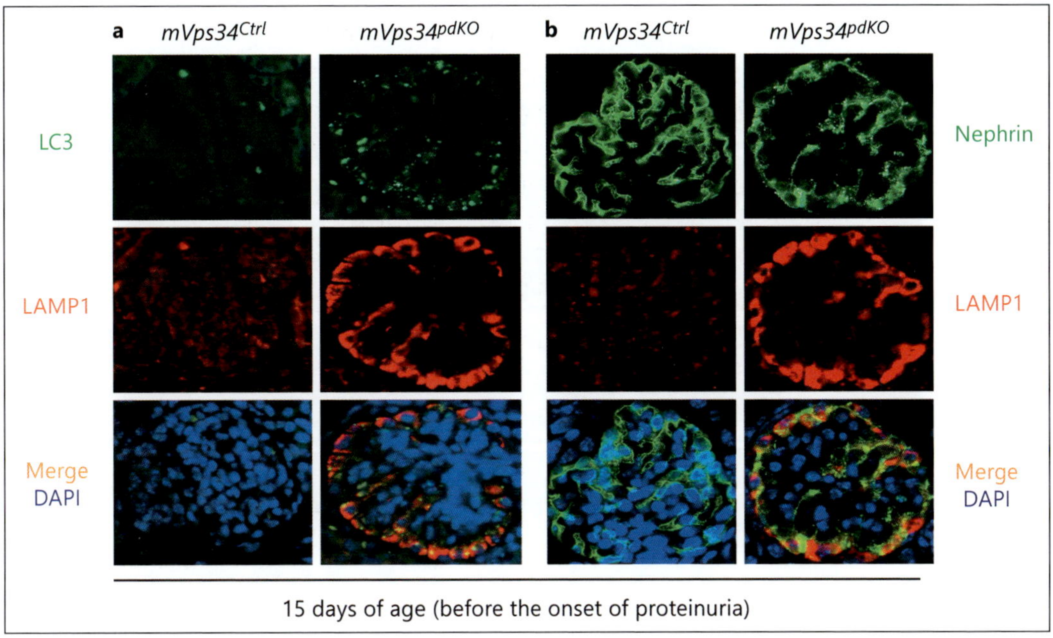

Fig. 3. Podocyte-specific mVps34 deletion increases the lysosome-associated membrane protein 1 (LAMP1) and the microtubule-associated protein 1 light chain 3 (LC3) specifically in the glomerular podocytes. **a** Immunofluorescence staining reveals that compared to their wild-type littermate control (*mVps34*^{Ctrl}) mice, podocyte-specific mVps34 knockout (*mVps34*^{pdKO}) mice exhibit markedly elevated levels of both LC3 punctate dots and LAMP1, with significant colocalization of the increased LC3 puncta with LAMP1, suggesting increased formation of autolysosomes, whereas other non-colocalized LC3 puncta indicates an increased number of autophagosomes. Note that the overall staining patterns of LC3 puncta and LAMP1 are very similar, suggesting that these increases are in the same cells. **b** Triple immunofluorescence staining confirmed that the increased LAMP1 staining in *mVps34*^{pdKO} mice are indeed confined to the nephrin-positive cells (podocytes). DAPI, 4′,6-diamidino-2-phenylindole. These images were derived from a copy of the figures 7C and 8B in the recent publication by Chen et al. [35], with the permission of *JASN*.

increased autophagy [34, 35]. Triple immunofluorescence staining localized markedly increased LC3-positive puncta in the *mVps34*-deleted podocytes, indicating increased autophagosomes (fig. 3). Electron microscopy confirmed that deletion of *mVps34* in podocytes indeed increases the number of double-membrane autophagosomes by 15 days of age before development of detectable proteinuria (fig. 4). Interesting, some of the increased LC3-positive punctate dots are colocalized with the increased lysosome-associated membrane protein 1 (LAMP1), indicating increased autolysosomes, while other LC3 puncta are not localized with LAMP1, representing increased autophago-

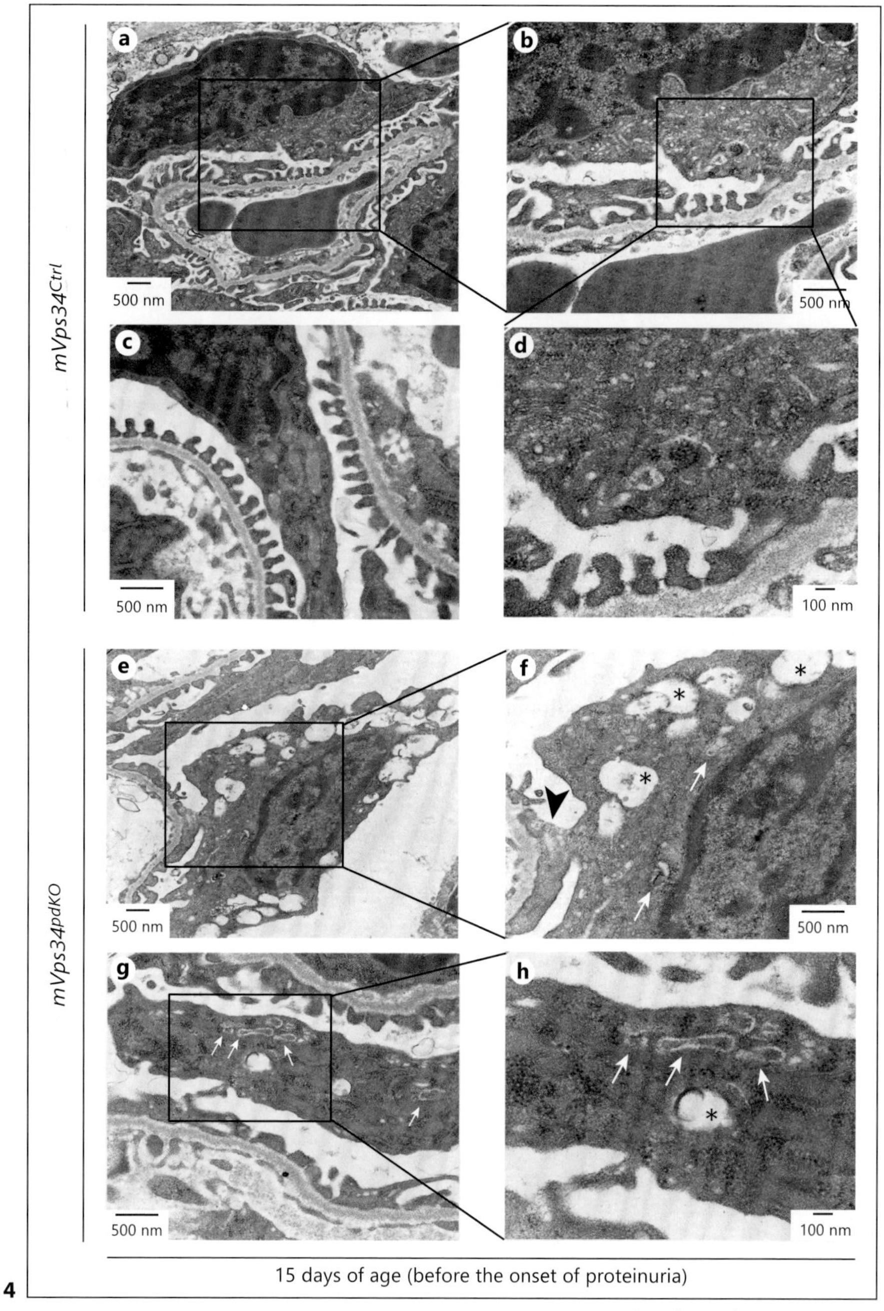

(*For legend see next page.*)

somes (fig. 3a). In addition, these *mVps34*-deleted podocytes also contain many large single-membrane vacuoles, indicative of enlarged endosomes (fig. 4). Of note, these accumulated intracellular vesicles continue to enlarge so they become readily visible and exhibit the striking podocyte vacuolization phenotype even by light microscopy when the mice reach 5–6 weeks of age [34, 35]. The enlarged vacuoles are stained strongly positive for both LAMP1 and LAMP2, indicating that the enlarged vacuoles originated from lysosomes [34, 35]. At this older age, the knockout mice have accumulated an increased number of double-membraned cytoplasmic vesicles that are characteristic of autophagosomes but have irregular inner membranes, as revealed by higher magnification of electron microscopy [34, 35]. These aberrant autophagosomes may be explained by the fact that mVps34-dependent production of phosphatidylinositol-3-phosphate is enriched on the inner membranes of autophagosomes. Impaired autophagic flux, however, does not appear to be the sole cause of early podocyte degeneration because podocyte-specific *Atg5* knockout mice exhibit a rather late-onset phenotype [18]. Furthermore, double knockout of *Vps34* and *Atg5* in podocytes results in a phenotype similar to that caused by a single knockout of *Vps34* [34]. These findings indicate that the early podocyte degeneration in *Vps34* knockout mice is only in part caused by defective autophagy. Further studies in Vps34-deficient mice as well as in Vps34 knockdown Garland cell nephrocytes, the podocyte-like cells of *Drosophila melanogaster*, revealed that endosomal trafficking, receptor-mediated endocytosis and maturation of early to late endosomes are all blocked in Vps34-deficient cells, confirming the important role of Vps34 in regulating these events in addition to autophagy in podocytes [34, 35]. Given the severe phenotype of podocyte-specific *Vps34* knockout mice, endocytosis, vesicle trafficking and an efficient autophagosomal-lysosomal degradation system seem to be a fundamental part of podocyte physiology [34, 36]. To this end, it should be pointed out again that in the podocyte-specific *Atg5* knockout mice, the Atg5/Atg7-independent alternative autophagic pathway, in which the lip-

Fig. 4. Podocyte-specific mVps34 deletion disrupts the intracellular vesicle trafficking and processing. In comparison with the wild type littermate controls (*mVps34*^{Ctrl}; **a–d**), podocyte-specific mVps34 knockout mice (*mVps34*^{pdKO}) develop numerous translucent vacuoles in their glomerular podocytes, representatively indicated by black asterisks (**e**, **f**), as well as an increased number of autophagosomes, pointed out by small white arrows (**f– h**) by 15 days of age before the onset of proteinuria, with mild focal foot process effacement being occasionally seen, as indicated by a black arrowhead (**f**). A scale bar is shown in each corresponding micrograph to indicate the magnification. These micrographs of electron microscopy were a copy of figure 9 in the recent publication by Chen et al. [35] with the permission of *JASN*.

idation of LC3-I to LC3-II that is generally believed to be a good indicator of autophagy does not occur, might come into action [19]. However, the lysosome is an absolutely indispensable player in any autophagic pathway so it is conceivable that lysosomal dysfunction in mouse as well as human podocytes would result in podocyte injury and glomerulosclerosis.

The prorenin receptor (PRR), also known as ATP6AP2 (ATPase, H^+-transporting, lysosomal accessory protein 2), is an accessory subunit of the vacuolar-type H^+-ATPase (V-ATPase) and highly expressed in podocytes [37]. V-ATPases play a critical role in the control of cellular and intracellular vesicle pH, which involves in numerous cellular processes including protein sorting, membrane fusion and trafficking, receptor-mediated endocytosis, and lysosomal protein degradation. The association of V-ATPases with PRR is required for its function [38]. Two groups have recently demonstrated that PRR is essential for podocyte survival by maintaining autophagy and protein turnover machinery [39, 40]. These studies suggest a functional connection between PRR, vacuolar acidification, and malfunction of autophagy. Podocyte-specific *PRR* deletion impairs vacuolar acidification, leading to defective autophagy, cytoskeleton reorganization, and massive accumulation of damaged proteins and organelles. These detrimental alterations induce podocyte degeneration, injury and cell death [39, 40]. Podocyte-specific *PRR* knockout mice died of early-onset severe proteinuria and acute kidney injury within 2–4 weeks after birth, which resembles the phenotype observed in podocyte-specific *mVps34* knockout mice [34, 35]. In addition, to complete lysosomal failure, impairments in protein sorting, vesicle trafficking, and receptor-mediated endocytosis might also cause this accelerated phenotype [38].

Several human LSDs are associated with an accumulation of storage products in podocytes, specifically leading to podocyte damage, proteinuria and renal failure [5]. A typical example is Fabry disease that is caused by the deficiency of lysosomal α-galactosidase A. Fabry disease is associated with dysfunction of many cell types and includes a systemic vasculopathy [41]. The histological characteristics are lysosomal inclusion bodies mainly deposited in podocytes in the kidney [42]. LC3, p62 and ubiquitin accumulate in glomerular cells in renal biopsy samples and in fibroblast and lymphoblast cell lines derived from patients with Fabry disease, suggesting an impaired autophagosome maturation and autophagic flux in this disorder [42]. A more recent study examined autophagy in an in vitro model of Fabry disease. Knockdown of α-galactosidase A in immortalized human podocytes leads to an accumulation of globotriaosylceramide (Gb3), accompanied by a dysregulated autophagy and a sustained decrease of mTOR and AKT signaling. These changes might predispose Fabry podocytes to further cellular damage [43].

mTOR Signaling in Autophagy Regulation in Podocytes

mTOR is a serine/threonine kinase that forms two distinct multiprotein complexes: mTORC1 and mTORC2. These two complexes phosphorylate different substrates and regulate a variety of essential cellular processes such as translation, transcription and autophagy. Recent genetic studies manipulating mTORC1 activity specifically in podocytes suggest that mTORC1 is a master regulator of podocyte development, maintenance and disease [44].

Compared to other glomerular cells, podocytes display both high levels of basal autophagy and mTORC1 activity, which is seemingly contradictory to the concept that mTORC1 negatively regulates autophagy. However, these interesting observations may suggest the existence of a unique mechanism involving the cooperation of mTORC1 and autophagy in podocytes. A recent study identified a special cytoplasmic compartment termed TOR-autophagy spatial coupling compartment (TASCC) in podocytes. The TASCC is a distinct cytoplasmic compartment located at the trans side of the Golgi apparatus (GA). The TASCC is mostly occupied by mTORC1, lysosomes, and autolysosomes but largely excludes developing autophagosomes (fig. 5). Given that mTORC1 is attracted by enriched lysosomes and autolysosomes, TASCC sequestration results in the formation of a spatial mTORC1 gradient within the cell, thereby allowing simultaneous activation of both mTORC1 and autophagy (fig. 5) [45]. In starvation-induced autophagy in normal rat kidney cells, mTOR is inhibited during initiation of autophagy. Following prolonged starvation, mTOR is reactivated and this event is autophagy-dependent and requires the degradation of autolysosomal products (autophagic lysosomal reformation). mTOR reactivation may in turn suppress autophagy and recycle lysosomes [46]. This feedback regulation is very important for a balance and fine-tuning between mTOR pathway and autophagy-lysosome pathway. Importantly, the spatial integration of rER-GA, autolysosomes, and mTORC1 also creates a self-regulating system, in which autolysosome-derived amino acids reinforce mTOR enrichment and activity through autophagic lysosomal reformation [45].

Mice with podocyte-specific *mTOR* knockout develop severe proteinuria at 3 weeks and renal failure by 5 weeks after birth. *mTOR*-deficient podocytes display an increased accumulation of LC3-II, autophagosomes, autolysosomes, and damaged mitochondria. Consistently, inhibition of mTOR in human podocytes by rapamycin also induces an accumulation of autophagosomes and autolysosomes [47]. These results suggest impaired autophagic flux in mTOR-deficient podocytes. Of note, mTOR deficiency leads to enhanced formation of autophagosomes as well as the accumulation of autolysosomes due to failure of the mTOR-dependent feedback regulation in these podocytes.

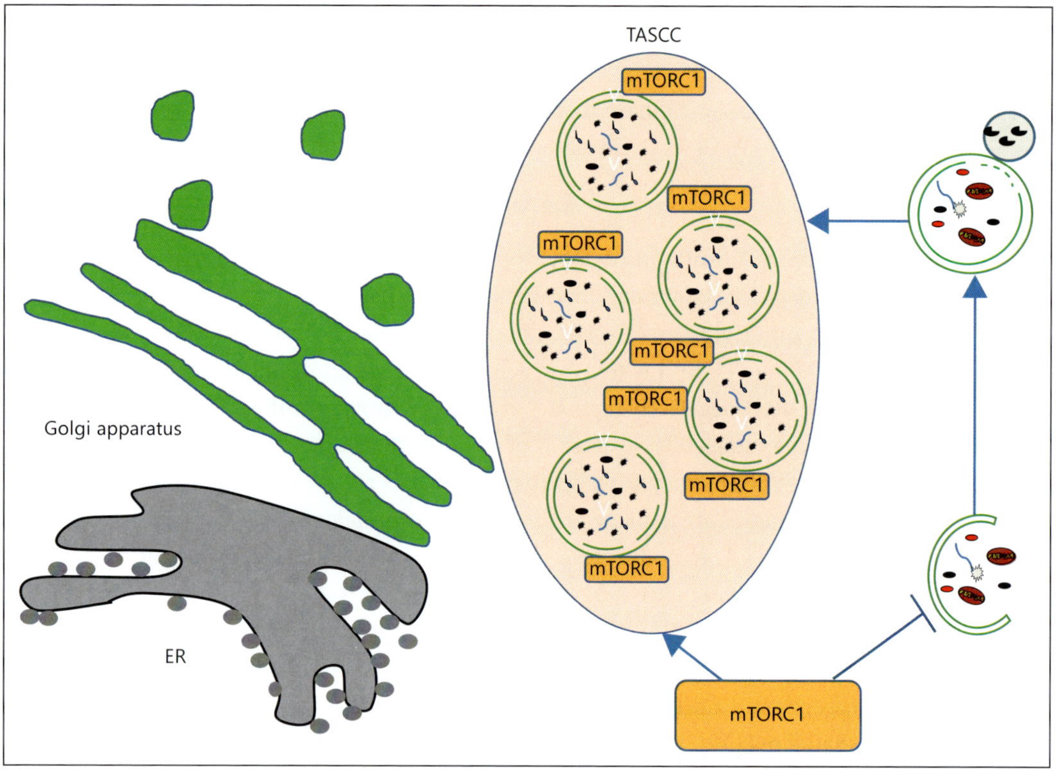

Fig. 5. Spatial regulation of mTORC1 and autophagy in podocytes. The TASCC is a distinct cytoplasmic compartment located at the trans side of the Golgi apparatus. The TASCC is mostly occupied by mTORC1, lysosomes, and autolysosomes but largely excludes developing autophagosomes. Sequestration of mTORC1 in TASCC by enriched lysosomes and autolysosomes results in the formation of a spatial mTORC1 gradient within the cell, thereby allowing simultaneous activation of both mTORC1 and autophagy.

Conclusions and Perspectives

Recent research has highlighted an essential role of autophagy in the homeostasis and function of podocytes. In these highly specialized cells, autophagy appears to be an important adaptive and cytoprotective mechanism in both physiological and pathological conditions. Future work will need to further elucidate the regulation of autophagy in podocytes in the context of aging and kidney diseases such as diabetic nephropathy, focal segmental glomerulosclerosis, glomerulonephritis and LSDs. Moreover, autophagy in podocytes as a potential therapeutic target for the prevention and treatment of kidney diseases needs to be evaluated.

Acknowledgements

The work was supported in part by grants from the National Basic Research Program of China 973 Program No. 2012CB517600, the National Natural Science Foundation of China (81370791), and the National Institutes of Health and Department of Veterans Administration of USA.

Disclosure Statement

The authors have no conflicts of interest to disclose.

References

1 Pavenstadt H, Kriz W, Kretzler M: Cell biology of the glomerular podocyte. Physiol Rev 2003;83:253–307.
2 Mundel P, Shankland SJ: Podocyte biology and response to injury. J Am Soc Nephrol 2002;13:3005–3015.
3 Wiggins RC: The spectrum of podocytopathies: a unifying view of glomerular diseases. Kidney Int 2007;71:1205–1214.
4 Tharaux PL, Huber TB: How many ways can a podocyte die? Semin Nephrol 2012;32:394–404.
5 Huber TB, Edelstein CL, Hartleben B, et al: Emerging role of autophagy in kidney function, diseases and aging. Autophagy 2012;8: 1009–1031.
6 Yang Z, Klionsky DJ: Eaten alive: a history of macroautophagy. Nat Cell Biol 2010;12:814–822.
7 Jin M, Liu X, Klionsky DJ: SnapShot: selective autophagy. Cell 2013;152:368–368 e2.
8 Mizushima N, Komatsu M: Autophagy: renovation of cells and tissues. Cell 2011;147:728–741.
9 Mizushima N, Yoshimori T, Ohsumi Y: The role of Atg proteins in autophagosome formation. Annu Rev Cell Dev Biol 2011;27: 107–132.
10 He C, Klionsky DJ: Regulation mechanisms and signaling pathways of autophagy. Annu Rev Genet 2009;43:67–93.
11 Kroemer G, Marino G, Levine B: Autophagy and the integrated stress response. Mol Cell 2010;40:280–293.
12 Ravikumar B, Sarkar S, Davies JE, et al: Regulation of mammalian autophagy in physiology and pathophysiology. Physiol Rev 2010; 90:1383–1435.
13 Levine B, Kroemer G: Autophagy in the pathogenesis of disease. Cell 2008;132:27–42.
14 Mizushima N, Levine B: Autophagy in mammalian development and differentiation. Nat Cell Biol 2010;12:823–830.
15 Choi AM, Ryter SW, Levine B: Autophagy in human health and disease. N Engl J Med 2013;368:651–662.
16 Asanuma K, Tanida I, Shirato I, et al: MAP-LC3, a promising autophagosomal marker, is processed during the differentiation and recovery of podocytes from PAN nephrosis. FASEB J 2003;17:1165–1167.
17 Mizushima N, Yamamoto A, Matsui M, et al: In vivo analysis of autophagy in response to nutrient starvation using transgenic mice expressing a fluorescent autophagosome marker. Mol Biol Cell 2004;15:1101–1111.
18 Hartleben B, Godel M, Meyer-Schwesinger C, et al: Autophagy influences glomerular disease susceptibility and maintains podocyte homeostasis in aging mice. J Clin Invest 2010; 120:1084–1096.
19 Nishida Y, Arakawa S, Fujitani K, et al: Discovery of Atg5/Atg7-independent alternative macroautophagy. Nature 2009;461:654–658.
20 Komatsu M, Waguri S, Ueno T, et al: Impairment of starvation-induced and constitutive autophagy in Atg7-deficient mice. J Cell Biol 2005;169:425–434.
21 Hara T, Nakamura K, Matsui M, et al: Suppression of basal autophagy in neural cells causes neurodegenerative disease in mice. Nature 2006;441:885–889.

PPAR-γ and Inflammation in Podocytes

Inflammation plays a significant role in the development and progression of various glomerular diseases. PPAR-γ agonists protect podocytes from puromycin aminonucleoside (PAN) or high-glucose associated injury, at least partially through suppressing inflammation in glomeruli. In a PAN-induced podocyte injury model [15, 16], PPAR-γ agonists remarkably decreased infiltrating macrophages, and decreased expression of plasminogen activator inhibitor-1 (PAI-1), which is the major inhibitor of tissue-type and urokinase-type plasminogen activators and prevents not only fibrinolysis but also proteolysis. PAI-1 is linked to increased extracellular matrix (ECM) accumulation and progressive glomerulosclerosis. In this model, vascular cell adhesion molecules 1 (VCAM-1) and intercellular adhesion molecules 1 (ICAM-1) released from endothelium in glomeruli may be involved in inhibition of inflammation by PPAR-γ [16]. In addition, PPAR-γ can reduce production of a cluster stimulating factors, and thus suppress secretion of cytokines such as TNF-α, IL-1β, IL-6. These effects ameliorate inflammatory reaction in podocytes and maintain morphology and function of podocytes.

PPAR-γ and Nephrin in Podocytes

The slit diaphragm is a cell-cell junction that shares features of tight and adherens junctions and displays anatomical regularity, with a gap thought to restrict the passage of proteins any larger than albumin. The gap is cross-linked by extracellular proteins, of which the transmembrane protein nephrin forms a major component. Nephrin is the key structural and signalling molecule within the slit diaphragm [11]. It is an integral podocyte membrane component, consisting of a large extracellular portion, which interacts with nephrin and NEPH1 molecules from adjacent foot processes, forming a zipper-like structure within the slit diaphragm. Human single gene disorders of nephrin (NPHS1) cause congenital/familial nephrotic syndromes, which were originally found in cohorts of Finnish patients. In several different types of human glomerular disease and animal models, nephrin expression has been shown to be downregulated or mislocalized in the podocyte [9, 11]. Therefore, restoration of nephrin production could be a mechanism to reduce glomerular filtration barrier dysfunction in proteinuric states.

Studies have shown that PPAR-γ agonists induce nephrin expression. Pioglitazone, which has been shown to prevent podocyte injury in several animal models (such as diabetic nephropathy, PAN, angiotensin II, aldosterone-induced podocyte injury), has also been shown to restore nephrin mRNA and protein levels in a Heymann nephritis model [17]. PPAR-γ agonists also prevented the serum deprivation or stretch-induced downregulation of nephrin in human podocytes [18, 19]. Indeed, peroxisome proliferator-responsive elements

(PPREs) in the promoter region of the nephrin gene have been identified, indicating NPHS1 as a putative PPAR target gene [17].

Besides maintaining integrity of filtering barrier, nephrin plays an important role in insulin signaling in podocytes. A key study showed that TZDs provide better renoprotection than insulin in an obese, hypertensive type 2 diabetic rat model, despite the TZD group of animals having poorer glycemic control and greater obesity [20]. Pioglitazone has been shown to increase nephrin transcription in transfected cells via a specific PPRE on the nephrin promoter, which may have relevance to insulin signaling pathways in the podocyte, as a specific mechanism of sensitizing podocytes to insulin. An alternative mechanism could be by direct structural benefits on glomerular cells, as suggested by the finding that pioglitazone reversed deranged gene expression pathways in glomeruli of db/db mice. The podocyte is thus considered to be the principle target of PPAR-γ, since TZDs, as insulin sensitizers, directly modulate the podocyte.

PPAR-γ and Mitochondria in Podocytes

Podocytes have many complex foot processes, which play important roles for maintaining the podocyte structure and functions. A considerable number of mitochondrial sections are observed in the cell body and narrow peripheral foot processes [9]. Morphological changes of mitochondria are exclusively detected in podocytes in FSGS model mice, mitochondrial cytopathy patients, and human FSGS patients caused by pamidronate [21]. Mitochondrial dysfunction represents a malfunction in biochemical processes, such as respiratory chain malfunction, or increased reactive oxygen species (ROS), and/or reduced copy number of mitochondria DNA (mtDNA) or mutation, resulting from either its own mtDNA or from nuclear DNA. The change of mitochondrial biogenesis might be one of important and initial steps of glomerular changes particularly podocyte injury in these conditions.

The mineralocorticoid aldosterone is now recognized as an important participant in the development and progression of kidney disease that worsens glomerular diseases through podocyte injury. A series in vivo and in vitro studies [22] demonstrated that aldosterone induced proteinuria and podocyte injury, accompanied by mitochondrial dysfunction, as assessed by the measurements of mitochondrial membrane potential, ATP levels, and mtDNA copy number, and the activity of complexes I, III, and IV. In cultured podocytes exposed to aldosterone, striking mitochondrial dysfunction, such as reduction of mitochondrial membrane potential, ATP levels, and mtDNA copy number were detected. Furthermore, the mineralocorticoid receptor antagonist significantly inhibited aldosterone-induced mitochondrial dysfunction and rescue the expression of nephrin [22]. In vitro studies further provide compelling evidence that dysfunction of mitochondrial is directly responsible for aldosterone-associated ROS generation and podocyte damage. What is important is that PPAR-γ agonist improves mitochondrial function. In vivo evidence demonstrated that rosiglitazone attenuated podocyte damage and proteinuria, accompanied by improvement of mitochondrial structure and function. A PPAR-γ siRNA, which knocks down the constitutively expressed PPAR-γ by >75%, abrogated the protective action of rosiglitazone on podocyte. Overexpression of PPAR-γ without concurrent rosiglitazone treatment elicited a protective effect in a manner resembling rosiglitazone in podocytes. The findings indicate that the effect of rosiglitazone on mitochondrial is PPAR-γ-dependent. It is also noted that although aldosterone induced PPAR-γ-coactivator-1α (PGC-1α) suppression, mitochondrial damage, and cellular injuries in podocytes, (SIRT1 a silent mating type information regulation 2 homolog 1) activator could attenuate mitochondrial malfunction and podocyte damages by increasing PGC-1 expression [23]. Mitochondrial energy metabolism by PGC-1 may also play important roles in podocyte functions.

Mitochondrial energy metabolism is likely required to maintain podocyte homeostasis. The control of podocyte energy metabolism seems to be a possible alternative strategy for attenuating podocyte injury. Therefore, PPAR-γ agonists can ameliorate glomerular diseases through the upregulation of mitochondrial function and maintenance of mitochondrial morphology.

PPAR-γ and Podocyte Apoptosis

PAN-induced podocyte injury was associated with apoptosis and decreased PPAR-γ mRNA expression. TZDs improved PAN-induced podocyte injury and restored podocyte differentiation, accompanied with upregulation of PPAR-γ mRNA and activity. In parallel, PPAR agonists increased cyclin-dependent kinase inhibitor p27 and the anti-apoptotic molecule Bcl-xL and decreased the pro-apoptotic caspase-3 activity [24]. Therefore, the protective effect of PPAR agonists in podocytes may be mediated in part by inhibition of apoptosis.

In rats, renal I/R significantly affects podocytes, causing effacement cytoplasmic edema, detachment, and cell loss. PPAR agonists rosiglitazone and ciglitazone reduce I/R-induced renal dysfunction and injury in Wistar rats, and pretreatment of rats with troglitazone decreases cell apoptosis in I/R renal injury [25]. In an in vitro model of oxygen/glucose deprivation-reoxygenation and/or serum deprivation, PPAR-γ agonists decreased cell loss mainly through an anti-apoptotic effect which was accompanied by the preservation of the Bcl-2 and Bax levels, and by the attenuation of caspase-3 activation [19]. However, whether these events result from a cross-talk between PPAR-γ and other signalling pathways (MAPK, NF-κB, etc.) and/or to binding to a PPRE in the promoter regions of PPAR-γ target genes, as reported for the bcl-2 gene [26], remain to be established. Interestingly, the finding that PPAR-γ agonists in the experiment restored expression of nephrin and that VEGF secreted by podocytes prevents phosphorylation of nephrin suggest a role of nephrin in podocyte apoptosis.

PPAR-γ and Microvascular Loops in Glomeruli

VEGF and angiopoietin are proteins secreted by podocytes that are necessary for survival of endothelial cells, podocytes, and mesangial cells. VEGF regulates slit diaphragm signaling and podocyte shape via VEGF-receptor 2-nephrin-nck-actin interactions [27]. VEGF, expressed by podocytes, activates VEGFR-2 on glomerular capillary endothelial cells. Pharmacological inhibition or targeted heterozygous deletion of VEGF in podocytes results in renal pathology manifested by loss of endothelial fenestrations in glomerular capillaries, proliferation of glomerular endothelial cells, loss of podocytes, and proteinuria in mice.

Endothelial damage is recently considered to be an important component of various sclerosing glomerular diseases, in addition to the well-known role of podocyte injury. In PAN model [15, 16] decreased endothelial cells correlated with sclerosis and glomerular capillary loss, whereas PPAR-γ agonist treatment was associated with increased number of endothelial cells. Possible mechanisms include decreased severity of initial injury, or subsequent repair and angiogenesis. As in other cells (e.g. smooth muscle cells, endothelial cells), PPAR agonists increased VEGF expression, it is thus possible that PPAR agonist increases podocyte VEGF synthesis and stimulates angiogenic progenitor cells differentiating to endothelial cells [15, 16]. The PPAR angiopoietin-related gene, also called Aglp4, a target of PPAR-γ, is an inhibitor of angiogenesis. Both VEGF-induced angiogenesis in vivo and vascular leakiness were significantly inhibited by Aglp4. PPAR-γ agonist treatment in PAN model was associated with decreased expression of Aglp4 [15, 16]. Therefore, PPAR agonist treatment protected glomerular capillaries both by increasing podocyte VEGF expression and by decreasing Aglp4, thus promoting open capillary loops.

Conclusion

Podocyte injury represents a common pathway leading to damaged glomerular filtration barrier and albuminuria in various types of CKD. Attenuating podocyte injury may offer a novel therapeutic intervention for this devastating disease. Accumulating in vitro and in vivo evidence demonstrates novel protective action of PPAR agonists against podocyte injury related to both diabetic and non-diabetic causes. The possible mechanisms include: (1) enhancement of insulin signaling leading to release of VEGF-A, (2) stimulation of nephrin expression, (3) improvement of mitochondrial biogenesis, (4) decreased apoptosis as a result of decreased caspase-3 and restoration of Bcl-xL, and (5) inhibition of inflammatory pathways, etc. PPAR agonists hold great potential as a therapeutic strategy in podocyte-related disease.

Acknowledgements

This work was supported by the National Basic Research Program of China 973 Program 2012CB517600 (No. 2012CB517602), Grant DK079162 from National Institute of Health, and Merit Review from Department of Veterans Affairs (to T.Y.). T.Y. is an Established Investigator from AHA and Research Career Scientist from the Department of Veterans Affairs.

Disclosure Statement

The authors have no conflicts of interest to disclose.

References

1 Yang J, Zhou Y, Guan Y: PPAR-γ as a therapeutic target in diabetic nephropathy and other renal diseases. Curr Opin Nephrol Hypertens 2012;21:97–105.

2 Ruan X, Zheng F, Guan Y: PPARs and the kidney in metabolic syndrome. Am J Physiol Renal Physiol 2008;294:F1032–F1047.

3 Yang J, Zhang D, Li J, Zhang X, Fan F, Guan Y: Role of PPAR-γ in renoprotection in type 2 diabetes: molecular mechanisms and therapeutic potential. Clin Sci (Lond) 2009;116: 17–26.

4 Lennon R, Welsh GI, Singh A, Satchell SC, Coward RJ, Tavare JM, Mathieson PW, Saleem MA: Rosiglitazone enhances glucose uptake in glomerular podocytes using the glucose transporter GLUT1. Diabetologia 2009; 52:1944–1952.

5 Coward RJ, Welsh GI, Yang J, Tasman C, Lennon R, Koziell A, Satchell S, Holman GD, Kerjaschki D, Tavare JM, Mathieson PW, Saleem MA: The human glomerular podocyte is a novel target for insulin action. Diabetes 2005;54:3095–3102.

6 Coward RJ, Foster RR, Patton D, Ni L, Lennon R, Bates DO, Harper SJ, Mathieson PW, Saleem MA: Nephrotic plasma alters slit diaphragm-dependent signaling and translocates nephrin, Podocin, and CD2 associated protein in cultured human podocytes. J Am Soc Nephrol 2005;16:629–637.

7 Coward RJ, Saleem MA: Podocytes as a target of insulin. Curr Diabetes Rev 2011;7:22–27.

8 Miyazaki Y, Cersosimo E, Triplitt C, DeFronzo RA: Rosiglitazone decreases albuminuria in type 2 diabetic patients. Kidney Int 2007; 72:1367–1373.

9 Imasawa T, Rossignol R: Podocyte energy metabolism and glomerular diseases. Int J Biochem Cell Biol 2013;45:2109–2118.

10 Welsh GI, Coward RJ: Podocytes, glucose and insulin. Curr Opin Nephrol Hypertens 2010;19:379–384.

11 Welsh GI, Hale LJ, Eremina V, Jeansson M, Maezawa Y, Lennon R, Pons DA, Owen RJ, Satchell SC, Miles MJ, Caunt CJ, McArdle CA, Pavenstadt H, Tavare JM, Herzenberg AM, Kahn CR, Mathieson PW, Quaggin SE, Saleem MA, Coward RJ: Insulin signaling to the glomerular podocyte is critical for normal kidney function. Cell Metab 2010;12:329–340.

12 Hale LJ, Hurcombe J, Lay A, Santamaria B, Valverde AM, Saleem MA, Mathieson PW, Welsh GI, Coward RJ: Insulin directly stimulates VEGF-A production in the glomerular podocyte. Am J Physiol Renal Physiol 2013; 305:F182–F188.

13 Coward RJ, Welsh GI, Koziell A, Hussain S, Lennon R, Ni L, Tavare JM, Mathieson PW, Saleem MA: Nephrin is critical for the action of insulin on human glomerular podocytes. Diabetes 2007;56:1127–1135.

14 Lennon R, Pons D, Sabin MA, Wei C, Shield JP, Coward RJ, Tavare JM, Mathieson PW, Saleem MA, Welsh GI: Saturated fatty acids induce insulin resistance in human podocytes: implications for diabetic nephropathy. Nephrol Dial Transplant 2009;24:3288–3296.

15 Yang HC, Ma LJ, Ma J, Fogo AB: Peroxisome proliferator-activated receptor-γ agonist is protective in podocyte injury-associated sclerosis. Kidney Int 2006;69:1756–1764.

16 Fogo AB: PPAR-γ and chronic kidney disease. Pediatr Nephrol 2011;26:347–351.

17 Benigni A, Zoja C, Tomasoni S, Campana M, Corna D, Zanchi C, Gagliardini E, Garofano E, Rottoli D, Ito T, Remuzzi G: Transcriptional regulation of nephrin gene by peroxisome proliferator-activated receptor-γ agonist: molecular mechanism of the antiproteinuric effect of pioglitazone. J Am Soc Nephrol 2006;17:1624–1632.

18 Miceli I, Burt D, Tarabra E, Camussi G, Perin PC, Gruden G: Stretch reduces nephrin expression via an angiotensin II-AT(1)-dependent mechanism in human podocytes: effect of rosiglitazone. Am J Physiol Renal Physiol 2010;298:F381–F390.

19 Miglio G, Rosa AC, Rattazzi L, Grange C, Collino M, Camussi G, Fantozzi R: The subtypes of peroxisome proliferator-activated receptors expressed by human podocytes and their role in decreasing podocyte injury. Br J Pharmacol 2011;162:111–125.

20 Ohtomo S, Izuhara Y, Takizawa S, Yamada N, Kakuta T, van Ypersele de Strihou C, Miyata T. Thiazolidinediones provide better renoprotection than insulin in an obese, hypertensive type 2 diabetic rat model. Kidney Int 2007;72:1512–1519.

21 Nagahama M, Sica DA: Pamidronate-induced kidney injury in a patient with metastatic breast cancer. Am J Med Sci 2009;338: 225–228.

22 Zhu C, Huang S, Yuan Y, Ding G, Chen R, Liu B, Yang T, Zhang A: Mitochondrial dysfunction mediates aldosterone-induced podocyte damage: a therapeutic target of PPAR-γ. Am J Pathol 2011;178:2020–2031.

23 Yuan Y, Huang S, Wang W, Wang Y, Zhang P, Zhu C, Ding G, Liu B, Yang T, Zhang A: Activation of peroxisome proliferator-activated receptor-γ coactivator-1α ameliorates mitochondrial dysfunction and protects podocytes from aldosterone-induced injury. Kidney Int 2012;82:771–789.

24 Kanjanabuch T, Ma LJ, Chen J, Pozzi A, Guan Y, Mundel P, Fogo AB: PPAR-γ agonist protects podocytes from injury. Kidney Int 2007;71:1232–1239.

25 Sivarajah A, Chatterjee PK, Patel NS, Todorovic Z, Hattori Y, Brown PA, Stewart KN, Mota-Filipe H, Cuzzocrea S, Thiemermann C: Agonists of peroxisome-proliferator activated receptor-γ reduce renal ischemia/reperfusion injury. Am J Nephrol 2003;23: 267–276.

26 Butts BD, Tran NL, Briehl MM: Identification of a functional peroxisome proliferator activated receptor response element in the 3′ untranslated region of the human bcl-2 gene. Int J Oncol 2004;24:1305–1310.

27 Izzedine H, Massard C, Spano JP, Goldwasser F, Khayat D, Soria JC: VEGF signalling inhibition-induced proteinuria: mechanisms, significance and management. Eur J Cancer 2010;46:439–448.

Tianxin Yang, MD, PhD
University of Utah and Veterans Affairs Medical Center
Division of Nephrology and Hypertension
30N 1900E, RM 4C224, Salt Lake City, UT 84132 (USA)
E-Mail Tianxin.Yang@hsc.utah.edu

Liu Z-H, He JC (eds): Podocytopathy. Contrib Nephrol. Basel, Karger, 2014, vol 183, pp 112–121
DOI: 10.1159/000360038

Podocyte Injury in FSGS

Wen Yi Ding · Moin A. Saleem
Academic Renal Unit, University of Bristol, Bristol, UK

Abstract

Background: The podocyte is central to the glomerular filtration barrier and is particularly important in the disease process of focal segmental glomerulosclerosis (FSGS). The podocyte is injured in two main ways in idiopathic nephrotic syndrome. The first is by single gene disorders, such as *NPHS1*, *NPHS2*, *CD2AP* and *TRPC6*, which has led to a focus on slit-diaphragm (SD) proteins and their role in FSGS. The second type of injury is by so-called circulating factors, which target the podocyte and affect the same signalling pathways as those damaged by single gene defects, suggesting a final common damage pathway, and resulting in deranged podocyte motility. ***Summary:*** As much of our understanding on the role of the podocyte in FSGS has stemmed from the discovery of SD proteins, we will first delineate the role of SD proteins in the maintenance of the filtration barrier, how changes in SD proteins contribute to podocyte injury and their role in FSGS. We will describe our current understanding of how they communicate with intracellular signals, which is intricately linked with the maintenance of their structure, as well as podocyte dynamics and motility, which influence resistance and subsequently filtration across the glomerular filtration barrier. ***Key Messages:*** Disruptions of the SD proteins, and subsequently of the actin and microtubule cytoskeleton, are consistently seen across models of FSGS. Studies on podocytes have given us increased clarity on the pathogenesis of FSGS, where cytokine regulation, fibrogenesis, control of cell proliferation and death as well as changes in cell type appear to play key roles.

Focal segmental glomerulosclerosis (FSGS) is a well-characterized histopathological diagnosis based on findings of sclerotic changes affecting a proportion of glomeruli. FSGS is a cause of significant morbidity in children with renal disease, comprising about 14% of children on haemodialysis in the United States [1].

Clinically, it tends to correlate with steroid-resistant nephrotic syndrome. Many patients with FSGS will progress to end-stage renal failure and require transplant. This devastating disease has a recurrence rate of 30–50% post-transplant in the paediatric population [2, 3]. This raised the likelihood of a circulating factor, which has been elusive thus far.

The glomerular filtration barrier comprises endothelial cells, the glomerular basement membrane and podocytes. The normal functioning of these structures is imperative in moderating normal filtration in the glomerulus and kidney. Healthy podocytes are particularly key in preventing the heavy proteinuria seen in nephrotic syndrome. Podocytes are highly specialized differentiated structures that comprise a cell body with major processes, secondary processes, and foot processes. The foot processes interdigitate and between neighbouring foot processes is a filtration slit called the slit diaphragm (SD). Podocytes are traditionally thought to be modified epithelial cells [4]. However, they also have features of smooth muscle cells, which is important in enabling the contractility and motility that we now believe are also critical to their function.

Initial discoveries in genetic FSGS demonstrated the significance of the SD in nephrotic syndromes. For example, mutations in *NPHS1* and *NPHS2* (coding for nephrin and podocin respectively) have been shown to be causative for nephrotic syndrome in various age groups. Novel technologies in genetic techniques like next-generation sequencing [5] and decreasing costs in these techniques have allowed identification of further mutations. These have given us more clarity as to the underlying cellular pathways in podocytes involved in the pathogenesis of FSGS.

What are the underlying cellular mechanisms in the disease process of FSGS that might be able to tie in the histology of sclerotic glomeruli with the clinical triad of proteinuria, hypoalbuminaemia and oedema? Some of the major questions in FSGS include: (1) What triggers the cascade of events resulting in nephrosis and glomerulosclerosis? (2) What cellular changes occur in the podocyte that result in the proteinuria and can this be rescued with targeted medication? (3) Does the proteinuria itself trigger a different mechanism resulting in the laying down of extracellular matrix in the glomerulus to cause sclerosis independently from the initial proteinuria-generating mechanism? (4) Does the foot process effacement seen in MCNS precede the eventual sclerotic changes seen in FSGS, and might there be ways to halt the progression of disease given that a large proportion of children respond to steroids initially in nephrotic syndrome? It is evident that the podocyte is key to this complex puzzle. In studying the podocyte, many of the pieces of this puzzle have been uncovered. We will detail below our current understanding of the podocyte in FSGS.

Injury of the Podocyte Slit Diaphragm

The SD, which is located between the foot processes of adjacent podocytes, is a complex multiprotein structure. The SD came into the spotlight with the discovery of *NPHS1* encoding nephrin in 1998 [6] in the context of congenital nephrotic syndrome (Finnish type). Subsequently, many other mutations in familial/genetic nephrotic syndromes have been found to code for proteins in the SD [7].

Nephrin

NPHS1 mutations tend to cause the most profound phenotype with onset of heavy proteinuria in utero or the first few months of life [6]. Nephrin is a large transmembrane protein of the immunoglobulin family containing 1,241 amino acids. It has a complex hairpin like structure with 8 extracellular IgG-like domains, a fibronectin type 3-like domain and 8 intracellular tyrosine phosphorylation sites [8]. Nephrin and Neph1 (another SD protein) are phosphorylated by Fyn to regulate actin cytoskeleton remodelling [4].

In vitro studies of podocytes exposed to plasma from patients with recurrent FSGS post-transplant have demonstrated that diseased plasma causes nephrin, podocin and CD2AP to assume an intracellular position rather than their normal cell surface position [9]. Other studies of patients with nephrotic syndrome have been conflicting in that some describe localization of nephrin away from the SD while others have not found this [10].

Nephrin interacts with several proteins within the SD and is important in actin cytoskeleton regulation. Nephrin interacts with CD2AP at the SD, and CD2AP is known to associate with various actin cytoskeletal regulating proteins which we will discuss in more detail below [11]. Nephrin has interactions with IQ-containing GTPase-activating protein (IQGAP) [12], Nck [13], phospholipase Cγ1 (PLCγ1) [14], and PI3K (phosphoinositide 3-OH kinase) [15] which signal to the actin cytoskeleton via Rho GTPases (Rac1, RhoA and cdc42), PAK (p21-activated kinase), WASP or Arp2/3 signalling [4].

In addition to cytoskeleton regulation, nephrin appears to have a role in the regulation of inflammation and cell survival. Nephrin and CD2AP, via its interactions with PI3K, stimulates AKT signalling in podocytes [16]. The PI3K/AKT pathway is one of the main antiapoptotic cell survival pathways. NF-κB is a transcription factor known to mediate inflammatory and immune responses, along with regulating cell growth and survival. Removing inhibition of NF-κB in mice resulted in glomerulosclerosis and proteinuria while in vitro studies showed that nephrin downregulation in podocytes resulted in increased NF-κB activation, which led the authors to conclude that the presence of nephrin might discourage the evolution of glomerular disease [17].

Podocin

NPHS2, which codes for podocin, was first described in nephrotic syndrome in 2000 [18]. Podocin is a member of the stomatin protein family, and contains a lipid recognition motif that enables it to localize to lipid rafts in the SD. It colocalizes with nephrin and the actin cytoskeleton at the SD [19] and augments nephrin activity by nearly 40-fold in vitro [20]. Podocin also associates with calcium-regulating ion channel TRPC6 at the SD and it has been suggested that this might form a mechanosensing complex that transduces a signal resulting in ion flux [21].

Podocin-deficient mice had massive proteinuria at birth, died in the first 5 weeks of life, lacked a SD and had diffuse mesangial sclerosis [22]. This phenotype is incongruous with most known human mutations of *NPHS2* which tend to cause a milder phenotype with later onset disease. Subsequently, using *Cre-loxP* technology, podocin was inactivated in the adult mouse kidney and this produced FSGS lesions after 4 weeks [23]. Podocin is likely to play a role in glomerular differentiation and it could be speculated that different mutations could determine whether its role in development is affected.

CD2AP

CD2AP is an 80-kDa protein known to interact with nephrin and podocin [24]. CD2AP knockout mice die of renal failure at 6–7 weeks of life and histology showed mesangial hyperplasia with foot process effacement [25]. It has a role in regulating the actin cytoskeleton. CD2AP forms a complex with capping protein and cortactin, both of which play a role in actin assembly and cell migration, to promote Arp2/3 complex-mediated actin assembly and the formation of branched actin filaments [26]. Capping protein and cortactin also interact with Rac1 to regulate cell-cell contact and this interaction is dependent on the presence of CD2AP [27].

Lack of CD2AP has been shown to increase podocyte apoptosis, via increased TGF-β signalling and increased translocation of dendrin from the SD into cells. Dendrin acts downstream to promote expression of cathepsin L, which in turn led to degradation of actin-regulating proteins synaptopodin, dynamin and RhoA [28].

α-Actinin-4

This is another actin-regulating protein that is important is cell adhesion and *ACTN4* mutation has been described in autosomal dominant FSGS [29]. Point mutation in vitro (K255E) disrupts actin assembly by enabling formation of actin/α-actinin-4 aggregates which results in mislocalization of actin away from cell-cell junctions [30], while the mouse equivalent (K256E) mutation resulted in increased ubiquitination, increased endoplasmic reticulum stress, and exacerbated apoptosis [31].

TRPC6

TRPC6 is a calcium-regulating ion channel that appears to have a gain of function mutation, i.e. increased calcium influx in late-onset FSGS [32]. Nephrin also appears to be involved in podocyte calcium homeostasis via PLCγ1 [14]. The increased calcium signalling mediated by TRPC6 increases the activity of RhoA, one of the RhoGTPases, with a resultant decrease in cell migration [33].

Actin Cytoskeleton, Microtubules and Motility

Podocytes, like neurons, have a cytoskeletal structure with thick microtubule-based processes, intermediate filaments and thin actin-based filaments at the peripheries, all of which interact to modulate cell structure and motility. It is clear from the studies listed above that the regulation of the actin cytoskeleton is key to regulating the podocyte foot processes and any disruption results in a diseased phenotype. The Rho GTPases (RhoA, Rac1 and cdc42) are very important proteins in coordinating changes in the actin cytoskeleton to maintain cell-cell adhesion, motility and contractility [34]. Podocyte movement has been demonstrated directly using multiphoton imaging in vivo [35] and it has been postulated that changes in motility regulated by the Rho GTPases underlie the pathogenesis of FSGS [34]. Plasma from 10 patients with FSGS produced a consistent phosphorylation of vasodilator-stimulated phosphoprotein via protease-activated receptor and increased podocyte motility [36]. Overall, alterations in podocyte motility are considered to underlie foot process effacement, with changes in the leading edge of the cell and its interactions with the glomerular basement membrane [37, 38]. It is clear that podocyte motility is tightly regulated, with both hypermotility and hypomotility being seen in diseased states. Some genetic defects result in podocyte hypomotlity e.g. alpha-actinin-4 or TRPC6, while podocyte injury from FSGS disease plasma or PAN (puromycin aminonucleoside) result in hypermotility [4].

Formins are essential in the modelling of both actin and microtubule cytoskeletons [39]. *INF2* (inverted formin 2) mutations were found to be responsible for 17% of familial FSGS in one series and 6 out of 7 altered residues corresponded to the mDia1 domain which is known to associate with IQGAP1 [40]. Microtubules are not only important structurally, but also have roles in trafficking molecules. *WT1* (Wilms' tumour suppressor gene) is a transcriptional factor important in podocyte differentiation and specific mutations result in FSGS associated either with Deny-Drash syndrome or Frasier syndrome. Deny-Drash syndrome podocytes show changes in cell structure as well as misexpression of actin cytoskeleton-associated proteins [41]. In a transient model of nephrotic syndrome in vitro, WT1-interacting protein translocated to the nuclei, which was dependent on microtubule transport by microtubule motor dynein, and this translocation changed podocyte actin assembly [42].

Another important gene in podocyte differentiation is *LMX1B*, which codes for a homeobox transcription factor. *LMX1B* mutations are typically associated with nail-patella syndrome (defects in nails, joints, eyes and glomerulus), but have recently been identified in cases of isolated FSGS [43]. LMX1B also has a role in fully differentiated podocytes, and in vitro studies show that LMX1B-deficient cells resulted in stiffer, less motile cells with increased actin and increased mRNA of actin-associated proteins Abra and Arl4c [44].

More recently, the role of microRNAs in FSGS has been investigated, and particularly interesting was a study on miR-193a. miR-193a was serendipitously identified in a search for candidates for breast cancer and miR-193a overexpression resulted in decreased WT1 expression and downregulation of podocalyxin and nephrin, both of which regulate the actin cytoskeleton [45].

Disorders of Glycosylation and Sialylation

Glycosylation is a powerful tool for post-translational modification of proteins and governs many biological processes. Nephrin is known to have 10 potential sites for N-glycosylation, and pharmacological inhibition of this in vitro demonstrated that glycosylation is important in facilitating nephrin localization to the cell membrane [46]. *ALG1* mutations are recognized to cause FSGS [7] and this codes for chitobiosyldiphosphodolichol β-mannosyltransferase, which is an enzyme in the glycosylation pathway. This pathway is also involved in sialylation, and it was observed that mice with uridine diphospho-N-acetylglucosamine 2-epimerase/N-acetylmannosamine kinase mutation resulted in hyposialylated podocalyxin, segmental splitting of the glomerular basement membrane and podocyte foot process effacement. These mice were given N-acetylmannosamine which increased sialylation and improved survival [47]. Another study of angiopoietin-like-4 (Angptl4), which is upregulated in minimal change disease, demonstrated that transgenic mice with overexpression of Angptl4 in podocytes had nephrotic range proteinuria and that N-acetylmannosamine decreased this by 40% [48]. N-acetylmannosamine is currently undergoing phase I trials for hereditary inclusion body myopathy, and it will be interesting to see whether this might be a promising new intervention for nephrotic syndrome.

Podocyte Injury, Proteinuria and Podocyte Regeneration

When those with steroid-resistant nephrotic syndrome are biopsied early, often one does not see the changes of FSGS on histology. This has been replicated in vivo with mice models, e.g. in Cre-Lox *NPHS2* or *LMX1b* knockout mice where onset of proteinuria is present for several weeks prior to histological change [23, 44]. One question raised is whether the proteinuria itself is an independent factor that leads to podocyte injury and the progression to FSGS. One study found that albuminuria prevents differentiation by renal progenitor cells into podocytes and that retinoic acid supplementation in an adriamycin-induced nephropathy mouse model improved proteinuria, podocyte numbers and differentiation [49]. It was previously thought that there was no mechanism for regeneration of the podocytes in adults but it now appears that the parietal epithelial cell is in fact a stem cell that produces podocyte progenitor cells [50]. This is an exciting prospect but little is currently known about how these cells function in FSGS. Prednisolone is used to treat FSGS and is known to act directly on podocytes to reduce apoptosis and cytoskeleton rearrangement, but also has been shown in mice to increase numbers of podocytes and podocyte progenitors [51]. Human umbilical mesenchymal stem cells have been used in an adriamycin-induced mouse model of FSGS and have had effects on reducing proteinuria and improved renal histological appearance [52]. Although it is unclear what mechanisms were responsible for this effect, the authors suggested that the reduction in pro-inflammatory cytokines along with the increase in anti-inflammatory cytokines might play a role. TGF-β_1, which is typically a pro-inflammatory cytokine, has fibrogenic effects via laminin, fibronectin and type IV collagen, has effects on cell growth/arrest and also stimulates epithelial-mesenchymal transition [53]. Clearly, there are complex interactions between cytokines and changes in cell types that underlie the progression to FSGS.

Conclusion

The podocyte has been integral to our understanding of FSGS and nephrotic syndrome. Genetic nephrotic syndromes have given us many clues leading to our understanding of the key roles of the SD proteins, the actin and microtubule cytoskeletons, cell motility and biochemical pathways like glycosylation and sialylation. The podocyte has also advanced our understanding of the role of cytokines, the possibility of cellular changes e.g. epithelial-mesenchymal transition in FSGS and that there may be a mechanism of podocyte regeneration via progenitor cells. We have not discussed the circulating factor in great detail here, as this will be discussed in the next chapter, but its effect is likely to be mediated via most of the actin regulatory pathways discussed here. This is a very exciting time for FSGS where the advancement in our understanding has potential in identifying new targets for therapy.

Disclosure Statement

The authors have no conflicts of interest to disclose.

References

1 Collaborative Studies NAPRTCS 2011 Annual Dialysis Report. Available from https://web.emmes.com/study/ped/annlrept/annlrept.html.

2 Ulinski T: Recurrence of focal segmental glomerulosclerosis after kidney transplantation: strategies and outcome. Curr Opin Organ Transplant 2010;15:628–632.

3 Fine RN: Recurrence of nephrotic syndrome/focal segmental glomerulosclerosis following renal transplantation in children. Pediatr Nephrol 2007;22:496–502.

4 Welsh GI, Saleem MA: The podocyte cytoskeleton – key to a functioning glomerulus in health and disease. Nat Rev Nephrol 2012;8:14–21.

5 McCarthy HJ, Bierzynska A, Wherlock M, Ognjanovic M, Kerecuk L, Hegde S, et al: Simultaneous sequencing of 24 genes associated with steroid-resistant nephrotic syndrome. Clin J Am Soc Nephrol 2013;8:1–12.

6 Kestilä M, Lenkkeri U, Männikkö M, Lamerdin J, McCready P, Putaala H, et al: Positionally cloned gene for a novel glomerular protein – nephrin – is mutated in congenital nephrotic syndrome. Mol Cell 1998;1:575–582.

7 McCarthy HJ, Saleem MA: Genetics in clinical practice: nephrotic and proteinuric syndromes. Nephron Exp Nephrol 2011;118:e1–e8.

8 Tryggvason K: Unraveling the mechanisms of glomerular ultrafiltration: nephrin, a key component of the slit diaphragm. J Am Soc Nephrol 1999;10:2440–2445.

9 Coward RJM, Foster RR, Patton D, Ni L, Lennon R, Bates DO, et al: Nephrotic plasma alters slit diaphragm-dependent signaling and translocates nephrin, podocin, and CD2 associated protein in cultured human podocytes. J Am Soc Nephrol 2005;16:629–637.

10 Welsh GI, Saleem MA: Nephrin-signature molecule of the glomerular podocyte? J Pathol 2010;220:328–337.

11 Shih NY, Li J, Cotran R, Mundel P, Miner JH, Shaw AS: CD2AP localizes to the slit diaphragm and binds to nephrin via a novel C-terminal domain. Am J Pathol 2001;159:2303–2308.

12 Deminie C, Rigothier C, Auguste P, Welsh GI, Mathieson PW, Saleem MA, et al: IQGAP1 interacts with components of the slit diaphragm complex in podocytes and is involved in podocyte migration and permeability in vitro. PLoS One 2012;7:e37695.

13 Jones N, Blasutig IM, Eremina V, Ruston JM, Bladt F, Li H, et al: Nck adaptor proteins link nephrin to the actin cytoskeleton of kidney podocytes. Nature 2006;440:818–823.

14 Harita Y, Kurihara H, Kosako H, Tezuka T, Sekine T, Igarashi T, et al: Phosphorylation of nephrin triggers Ca^{2+} signaling by recruitment and activation of phospholipase $C\gamma_1$. J Biol Chem 2009;284:8951–8962.

15 Zhu J, Sun N, Aoudjit L, Li H, Kawachi H, Lemay S, et al: Nephrin mediates actin reorganization via phosphoinositide 3-kinase in podocytes. Kidney Int 2008;73:556–566.

Liu Z-H, He JC (eds): Podocytopathy. Contrib Nephrol. Basel, Karger, 2014, vol 183, pp 122–131
DOI: 10.1159/000360048

Circulating Soluble Urokinase Receptor and Focal Segmental Glomerulosclerosis

Changli Wei · Jochen Reiser

Department of Internal Medicine, Rush University, Chicago, Ill., USA

Abstract

Background: Clinical observations of rapid proteinuria after kidney transplantation suggest the existence of a causative permeability factor in the circulation of focal segmental glomerulosclerosis (FSGS) patients. Searching for such factor(s) however has been extremely painstaking and extends over the last five decades. We recently identified soluble urokinase receptor (suPAR) as a circulating FSGS factor. ***Summary:*** suPAR represents the soluble form of a glycosylphosphatidylinositol-anchored membrane protein and has been implicated in a number of clinical conditions. With three different rodent models, we showed that acute induction of circulating suPAR could deposit into the glomerulus and then activate podocyte αVβ3 integrin activity; chronic suPAR overexpression caused kidney changes resembling FSGS, which could be ameliorated by suPAR blocking antibody or by expression of suPAR deficient in β3 integrin binding. Analysis of circulating suPAR in four large cohorts of primary FSGS and normal renal average function with more than 300 patients indicates that suPAR level is increased in the majority of FSGS patients. Additionally, circulating suPAR could be removed by plasmapheresis. Removal of suPAR and decreasing podocyte αVβ3 integrin activity is associated with proteinuria remission in recurrent FSGS. suPAR-induced podocyte β3 integrin was observed particularly with FSGS and is independent from inflammation. ***Conclusion:*** suPAR drives αVβ3 integrin signaling in podocytes during FSGS. ***Key Message:*** Circulating suPAR contributes to FSGS. Current ELISA measurements of serum suPAR are handicapped by suPAR's association with GFR and by the lack of detection for aberrant suPAR forms relevant to FSGS. The specificity of ELISA testing can be improved by analysis of the suPAR-αVβ3 integrin signaling axis using cultured human podocytes. The future holds the identification of specific pathological suPAR forms for FSGS and renewed ELISA testing to develop novel therapies of suPAR modification for patients.

Focal Segmental Glomerulosclerosis (FSGS)

FSGS is a glomerular disorder characterized by marked proteinuria, often coupled with steroid resistance, hypertension, and high incidence (40–60%) of progression to renal failure. It is a common and progressive glomerular disease in children and also accounts for 20–25% of idiopathic nephrotic syndrome in adults. The etiology of FSGS is defined by its histological pattern while its mechanism(s) is/are being heavily studied.

Evolution of the Circulating Permeability Factor Concept

The hypothesis of circulating factor in nephrotic syndrome was originally proposed by Shalhoub [1] in 1974, when he postulated that abnormality of T-cell function results in secretion of a circulating chemical mediator that damages the glomerular basement membrane. Over the last five decades or so, there were several factors involved in minimal change disease (MCD) surfaced including vascular permeability factor [2], hemopexin [3], CD87 [4], and angiopoietin-like-4 [5]. However, none of these factors have been either verified with a large cohort of MCD patients or provided sufficient underlying mechanisms. With regard to FSGS, since the first report of recurrence of FSGS after transplantation [6], there have been several observations that strongly support the hypothesis of a causative, permeability factor in the circulation of FSGS patients. This evidence includes (i) recurrent FSGS can be associated with massive proteinuria within hours of a kidney transplant; (ii) injection of FSGS plasma or its fractions into rats causes proteinuria; (iii) FSGS can be treated with plasmapheresis, presumably to remove renal injurious substance(s) from the circulation, and (iv) sera from patients with FSGS increase albumin permeability in isolated glomeruli and transient nephrotic syndrome has been transmitted to a newborn from a mother with FSGS [7]. These data suggest that primary FSGS can be a systemic disorder where a circulating factor is the pathogenic culprit causing podocytopathy.

Searching for the circulating FSGS factor however has been extremely painstaking and has long been considered as one of the holy grails of nephrology [8]. The pioneers, Savin and colleagues [9], have been studying permeability activity in the plasma of FSGS patients for more than 20 years with an in vitro functional permeability assay based on the isolated rat glomeruli. In 1999, they were successful in enriching plasma for a permeability factor using standard biochemical techniques [9]. Their studies suggested that the factor is a protein of molecular size between 20 and 50 kDa [9]. In 2008, cardiotrophin-like cytokine-1 (CLC-1) was proposed as a putative FSGS permeability factor [9]. However, it is not yet clear if and how CLC-1 causes FSGS and whether it can be confirmed in a large cohort of FSGS patients. Studies in our laboratory suggested that one major permeability factor in both native and recurrent FSGS is a soluble form of the urokinase-type plasminogen activator receptor (suPAR).

suPAR

suPAR represents the soluble form of the urokinase-type plasminogen activator receptor (uPAR), which is a glycosylphosphatidylinositol (GPI)-anchored three-domain, and heavily glycosylated membrane protein that can bind to several ligands including uPA, vitronectin or integrins [10]. suPAR stems from cleavage and release of the membrane-bound uPAR, and can be detected in plasma, urine, blood, serum, cerebrospinal fluid and saliva [11]. uPAR is probably most susceptible for cleavage at the linker region of D_I, connecting it with D_{II} and D_{III}. While cleavage from membrane-bound uPAR could contribute to the majority forms of suPAR (D_I, $D_{II}D_{III}$, $D_ID_{II}D_{III}$), rodents also express soluble uPAR that is being synthesized without the GPI anchor [11]. Whether this form also exists in humans is still unclear.

Evidence Indicating suPAR as a FSGS Circulating Factor

Preclinical Studies

uPAR present on multiple cells, including podocytes. We have originally shown that induction of membrane-bound uPAR on podocytes enhances $\alpha V\beta 3$ integrin expression in the kidney biopsies of FSGS patient and in several rodent proteinuria models [12]. Of relevance to FSGS circulating factor is that proteolytic cleavage of uPAR from cells such as neutrophils can release suPAR into circulation [13]. Subsequently, we showed that elevated levels of serum suPAR promote suPAR deposits in mouse kidneys along podocytes, which was associated with an increase in $\beta 3$ integrin activity [14]. Using multiple mouse models, we demonstrated that injection of suPAR in uPAR null mice as well as chronic overexpression of the suPAR (BC010309) in wild-type mice generated features of FSGS such as foot process effacement, proteinuria, and segmental lesions, all of which could be ameliorated with addition of uPAR-specific monoclonal antibody or by expression of suPAR deficient in $\beta 3$ integrin binding [14]. Using a hybrid mouse kidney transplantation model that has a uPAR null kidney and a wild-type kidney in place, we found that endogenously induced suPAR (using LPS injection) could circulate into the grafted uPAR null kidney and deposit into the glomerulus along podocytes. Taken together, these preclinical studies suggested that enhanced circulating suPAR induces changes in podocytes similar to that of primary FSGS via induction of the αVβ3 signaling pathway [14].

Multiple Large FSGS Cohort Studies
Using commercially available ELISA assay by R&D, we have shown that suPAR levels were increased in two thirds of a multicenter, biopsy-proven cohort of pediatric and adult FSGS patients, including both native and recurrent FSGS cases, but not in other analyzed proteinuric diseases including MCD, membranous nephropathy and preeclampsia. More recently, we have extended these findings by studying serum suPAR levels in a large number of patients with FSGS from two well-characterized cohorts from the North America FSGS clinical trial (FSGS-CT) (pediatric and adult patients, normal average GFR) and a Europe FSGS consortium, PodoNet (pediatric patients, normal average GFR). Using 3 ng/ml as a tentative cut-off value, we demonstrated that suPAR serum levels were elevated in 84 and 55% of patients, respectively. Importantly, the elevation of suPAR in FSGS patients was not related to inflammation because concurrent C-reactive protein levels were not elevated in these groups [15]. Consistent with our major findings, Huang et al. [16] showed elevation of suPAR in the majority of plasma samples from a biopsy-proven primary FSGS patient cohort in China, but not in membranous nephropathy or MCD.

Transmission of suPAR from a Mother with FSGS to Her Baby
In 2001, Kemper et al. [17] described a case in which a mother with FSGS gave birth to a girl who developed proteinuria at birth, which then lasted for several days. The newborn's proteinuria gradually decreased and eventually resolved completely. We analyzed the original blood samples from the mother and the baby girl for suPAR and compared them with samples from healthy newborns. We found a mean suPAR level of 2,884 pg/ml in normal newborns versus highly elevated levels in both the mother (4,635 pg/ml) and her baby (5,225 pg/ml) at the time of the infant's reported peak proteinuria [18]. The finding of increased suPAR level and transient proteinuria in an infant born to a mother with FSGS is consistent with a role of suPAR as circulating FSGS factor.

Involvement of suPAR in Recurrent FSGS
From the first multicenter FSGS cohort for suPAR study, we found the highest level of serum suPAR in pretransplantation blood from subjects with FSGS who later developed recurrent FSGS after transplantation. Comparatively, the pretransplantation serum suPAR in patients who did not suffer FSGS recurrence after transplantation was similar to that observed in native non-transplant FSGS

patients [14]. Thus the pretransplantation suPAR level may work as a predictor of heightened risk of recurrent FSGS after transplantation. We also compared suPAR concentrations in transplant FSGS patients 1 year after transplantation and found significantly higher serum suPAR in patients who developed recurrent FSGS than in those who did not recur [14]. The longitudinal analysis of serum suPAR level in subjects with recurrent and non-recurrent FSGS indicated that the subjects with recurrent FSGS had sustained higher suPAR serum concentrations, compared to those without any recurrence. Furthermore, in 62% subjects that developed recurrent FSGS, suPAR was further increased during the course of the 1-year interval, a finding that was in contrast to the suPAR concentrations we observed in subjects with no recurrence after kidney transplantation [14]. More recently, Franco Palacios et al. [19] suggested urinary suPAR as a possible pretransplantation biomarker for recurrent FSGS. Cleary, the role of urinary suPAR and how suPAR is processed in the kidney requires further study.

Plasmapheresis Removing suPAR from Circulation

Case and series study has shown that plasmapheresis is effective in up to 60–70% recurrent FSGS patients to achieve partial or complete remission [20]. For plasmapheresis to be applied, the molecule should be 15 kDa or larger. suPAR in most cases is a mixture of different cleaved or secreted fragments, ranging from 16 to 60 kDa by molecule mass [10]. Thus, in theory, plasmapheresis is justified for suPAR removal. Indeed, we observed that plasmapheresis significantly reduced suPAR serum concentrations in subjects with recurrent FSGS. We then studied the effects of before- and after-pheresis serum samples from subjects with FSGS on podocyte β3 integrin activity by measuring AP5 signal. We found that plasmapheresis significantly lowered podocyte β3 integrin activity when incubated podocytes with serum from subjects with recurrent FSGS [14]. Next, we studied 4 clinical cases of recurrent patients with FSGS who received plasmapheresis after transplantation. All patients had elevated suPAR serum concentrations before transplantation. After serial plasmapheresis treatments, 2 patients reached a clinical remission; their serum suPAR concentrations fell below 2,000 pg/ml, and, notably, their serum also lost the capacity to induce podocyte β3 integrin activity (the latter being the critical step to induce podocyte injury). In contrast, the other 2 patients remained in recurrence despite plasmapheresis. Their serum suPAR concentrations remained elevated and their sera still caused strong podocyte β3 integrin activity [14]. These findings suggest that the disease-stabilizing effects of plasmapheresis depend on lowering individual serum suPAR to levels that sharply decrease podocyte β3 integrin activity. In addition, we have tested the possibility of managing recurrent FSGS patients by apheresis therapy that is guided by serum suPAR levels. We followed up a 59-year-old female patient with biopsy-proven primary FSGS. She had highly elevated serum suPAR level at the day of transplantation. She suffered from delayed graft function and was found to have heavy proteinuria at initiation of diuresis. Repeat graft biopsies were consistent with recurrent FSGS. Since high suPAR values remained posttransplantation, plasmapheresis was initiated on postoperative day 37. Serum suPAR levels were successfully reduced by 42%, so was proteinuria, achieving partial remission. Attempts to wean plasmapheresis however failed due to severe oliguria following cessation of treatment. The patient was thus subsequently switched to immunoadsorption, which showed equal efficacy in suPAR reduction. In addition to serum suPAR monitoring, we evaluated the suPAR-induced downstream effect on human podocyte β3 integrin activation. We found that the suPAR reduction due to intensified apheresis reduced podocyte β3 integrin activation as measured by AP5 staining to a minimum of 1,145 mean fluorescence intensity, which is considered normal in this experimental setting. Moreover, the decrease in podocyte β3 integrin activation was associated with a significant decrease in proteinuria as well. Electron microscopy shows partial recovery of podocyte foot process pathology on postoperative day 437. This case study suggests that severe recurrent posttransplantation FSGS may be managed by an approach using apheresis therapy that is guided by serum suPAR levels [21].

suPAR and Podocyte Effacement

Podocyte effacement is the first pathological manifestation of FSGS after transplantation. In a retrospective study of 21 cases of recurrent FSGS, Alachkar et al. [22] found that circulating suPAR levels at time of diagnosis correlate with the severity of effacement. Furthermore, a complete or even partial response to plasmapheresis (alone or combined with rituximab) results in significant decrease in suPAR levels, resolution of foot process effacement and improvement in proteinuria.

Taken together, these reported preclinical and clinical studies indicate that suPAR is a circulating factor that drives FSGS.

The Challenge and Current Gap between Bench and Bedside

The identification of suPAR as a FSGS factor might be confusing since suPAR levels are also elevated in other diseases including cancer and infection that do not routinely present with proteinuria, and some patients develop recurrent FSGS but do not have high circulating suPAR levels [23]. A number of studies suggest that increased activation of the immune system leads to increased serum suPAR levels, which has been documented in several pathological conditions, including paroxysmal nocturnal hemoglobinuria, HIV-1 infection, pneumococcal bacteremia, malaria, sepsis, bacterial and viral CNS infection, tuberculosis, and also various forms of solid tumors [10]. Thus, it is assumed by some researchers that levels of suPAR reflect an overall systemic inflammation and immune activation that may be common to many life-threatening diseases. Therefore, it is not surprising that the concept of suPAR being the circulating factor that drives FSGS has been taken with some skepticism [23]. Particularly, some studies with much smaller size of FSGS patient samples could not validate the specificity of high suPAR levels for primary FSGS [24, 25]. Although the small size of these studies does not allow to draw firm conclusions, and clinical and histologic differentiation between primary FSGS and secondary FSGS can be cumbersome, these reports indeed question the clinical usefulness of high suPAR levels for diagnosis of primary FSGS in patients [26]. Most recently, more suPAR studies have been published presenting different and often conflicting data with each other, highlighting inconsistencies among the cohort and sampling conditions as well as questioning a uniform use of the current suPAR ELISA [27–31].

Suggested Answers and Future Work

Avoiding the Pitfalls in suPAR Measurement and Analysis

Given the complexity nature of FSGS, and the presence of suPAR in both healthy and diseases, the following factors should be taken into close account when measuring suPAR and assessing its significance: (a) *Assay platform*: Although it is not technically difficult to set up the ELISA-based assay, several factors could be easily overlooked as we found in our communication with colleagues working in this field. Before any reading can be reported, both inter- and intra-assay precision should be assessed with acceptable low coefficient of variation (5% or less). Standard curve should be generated from each assay and validated. (b) *Sample size and time of sampling*: Sample size should be large enough to cover the variations especially when a single-center cohort is analyzed. Time of sampling is also an issue easily overlooked. Samples obtained at different clinical phases (before or after plasmapheresis, before or after transplantation, change of medication, coexisting disease, etc.) should be matched for comparison. (c) *Serum or plasma*: While suPAR itself is relatively stable in blood circulation, its values differ in different forms of samples, with serum level usually higher than plasma level. Even with plasma, EDTA can generate different suPAR readings from heparin. Unfortunately, it is not uncommon in clinical studies that serum and plasma are mixed up. (d) *Confounding factors*: When evaluating the role of suPAR in FSGS or other glomerular disorders, one should be aware whether other conditions associated with high suPAR levels coexist, e.g. infections, tumors. With the former, concurrent analyzing serum C-reactive protein level will be important. (e) *End-stage kidney disease (ESKD)*: ESKD patients often have very limited or no renal function, retention of circulating suPAR could be significant, thus may contribute to increasing suPAR levels. As current suPAR test kits could not distinguish pathological or aberrant suPAR fragments, evaluation of suPAR-induced podocyte αVβ3 integrin activity can provide further answers and studies are well underway to address this issue.

Determining Podocyte αVβ3 Integrin Activity

We have found that enhanced circulating suPAR levels induce podocyte αVβ3 integrin activity in multiple animal models. This provides us the rationale to examine the αVβ3 integrin activity particularly when high serum suPAR levels occur in non-FSGS conditions. Although high suPAR levels are often associated with infections such as sepsis or septic shock, our studies indicate that the high suPAR-containing sera obtained from these infection patients did not induce podocyte β3 integrin activity, while the primary FSGS sera did [unpublished data]. In the same token, high circulating suPAR has also been observed in some non-FSGS ESKD patients [19] and podocyte β3 integrin activity failed to activate podocyte αVβ3 integrin [unpublished data]. These data further indicate that enhanced suPAR-β3 integrin signaling is specific largely to primary FSGS.

Pinpointing the Pathological Forms of suPAR

As discussed above, suPAR has multiple forms representing different fragments. We hypothesize that some pathological suPAR fragments but not total suPAR account for FSGS. Normally, suPAR is a heavily glycosylated protein. Our recent findings suggest that loss of glycosylation renders suPAR a much more potent activator of αVβ3 integrin [unpublished data]. It will be a priority to isolate these pathologically glycosylated suPAR fragments from FSGS patients and develop a suPAR ELISA that can specifically detect various suPAR fragments.

Conclusion

Circulating suPAR is increased in the majority of primary FSGS patients which is particularly relevant in patients with normal or mildly reduced GFR. Elevated suPAR-β3 integrin signaling contributes to the pathogenicity of primary and recurrent FSGS and its level can be manipulated through suPAR removal devices and potential biologics. While future work identifying the pathological suPAR fragments is the key in addressing the specificity of suPAR in different diseases, a removal trial using antibodies that eliminate 'all' suPAR forms will be paramount to test if suPAR is a culprit for FSGS in humans. Furthermore, suPAR's direct contribution to general human health needs to be further clarified.

References

1 Shalhoub RJ: Pathogenesis of lipoid nephrosis: a disorder of T-cell function. Lancet 1974; 2:556–560.

2 Lagrue G, Xheneumont S, Branellec A, Hirbec G, Weil B: A vascular permeability factor elaborated from lymphocytes. I. Demonstration in patients with nephrotic syndrome. Biomedicine 1975;23:37–40.

3 Bakker WW, van Dael CM, Pierik LJ, van Wijk JA, Nauta J, Borghuis T, Kapojos JJ: Altered activity of plasma hemopexin in patients with minimal change disease in relapse. Pediatr Nephrol 2005;20:1410–1415.

4 Garin EH, Diaz LN, Mu W, Wasserfall C, Araya C, Segal M, Johnson RJ: Urinary CD80 excretion increases in idiopathic minimal-change disease. J Am Soc Nephrol 2009;20: 260–266.

5 Clement LC, Avila-Casado C, Macé C, Soria E, Bakker WW, Kersten S, Chugh SS: Podocyte-secreted angiopoietin-like-4 mediates proteinuria in glucocorticoid-sensitive nephrotic syndrome. Nat Med 2011;17:117–122.

6 Hoyer JR, Vernier RL, Najarian JS, Raij L, Simmons RL, Michael AF: Recurrence of idiopathic nephrotic syndrome after renal transplantation. Lancet 1972;2:343–348.

7 Jefferson JA, Shankland SJ: Has the circulating permeability factor in primary FSGS been found? Kidney Int 2013;84:235–238.

8 Glassock RJ: Circulating permeability factors in the nephrotic syndrome: a fresh look at an old problem. J Am Soc Nephrol 2003;14:541–543.

9 McCarthy ET, Sharma M, Savin VJ: Circulating permeability factors in idiopathic nephrotic syndrome and focal segmental glomerulosclerosis. Clin J Am Soc Nephrol 2010; 5:2115–2121.

10 Thuno M, Macho B, Eugen-Olsen J: suPAR: the molecular crystal ball. Dis Markers 2009; 27:157–172.

11 Reiser J, Wei C, Tumlin J: Soluble urokinase receptor and focal segmental glomerulosclerosis. Curr Opin Nephrol Hypertens 2012;21: 428–432.

12 Wei C, Möller CC, Altintas MM, Li J, Schwarz K, Zacchigna S, Xie L, Henger A, Schmid H, Rastaldi MP, Cowan P, Kretzler M, Parrilla R, Bendayan M, Gupta V, Nikolic B, Kalluri R, Carmeliet P, Mundel P, Reiser J: Modification of kidney barrier function by the urokinase receptor. Nat Med 2008;14: 55–63.

13 Shankland SJ, Pollak MR: A suPAR circulating factor causes kidney disease. Nat Med 2011;17:926–927.

14 Wei C, El Hindi S, Li J, Fornoni A, Goes N, Sageshima J, Maiguel D, Karumanchi SA, Yap HK, Saleem M, Zhang Q, Nikolic B, Chaudhuri A, Daftarian P, Salido E, Torres A, Salifu M, Sarwal MM, Schaefer F, Morath C, Schwenger V, Zeier M, Gupta V, Roth D, Rastaldi MP, Burke G, Ruiz P, Reiser J: Circulating urokinase receptor as a cause of focal segmental glomerulosclerosis. Nat Med 2011; 17:952–960.

15 Wei C, Trachtman H, Li J, Dong C, Friedman AL, Gassman JJ, McMahan JL, Radeva M, Heil KM, Trautmann A, Anarat A, Emre S, Ghiggeri GM, Ozaltin F, Haffner D, Gipson DS, Kaskel F, Fischer DC, Schaefer F, Reiser J; PodoNet and FSGS CT Study Consortia: Circulating suPAR in two cohorts of primary FSGS. J Am Soc Nephrol 2012;23:2051–2059.

16 Huang J, Liu G, Zhang YM, Cui Z, Wang F, Liu XJ, Chu R, Chen Y, Zhao MH: Plasma soluble urokinase receptor levels are increased but do not distinguish primary from secondary focal segmental glomerulosclerosis. Kidney Int 2013;84:366–372.

17 Kemper MJ, Wolf G, Müller-Wiefel DE: Transmission of glomerular permeability factor from a mother to her child. N Engl J Med 2001;344:386–387.
18 Kemper MJ, Wei C, Reiser J: Transmission of glomerular permeability factor soluble urokinase plasminogen activator receptor (suPAR) from a mother to child. Am J Kidney Dis 2013;61:352.
19 Franco Palacios CR, Lieske JC, Wadei HM, Rule AD, Fervenza FC, Voskoboev N, Garovic VD, Zand L, Stegall MD, Cosio FG, Amer H: Urine but not serum soluble urokinase receptor (suPAR) may identify cases of recurrent FSGS in kidney transplant candidates. Transplantation 2013;96:394–399.
20 Cravedi P, Kopp JB, Remuzzi G: Recent progress in the pathophysiology and treatment of FSGS recurrence. Am J Transplant 2013;13: 266–274.
21 Morath C, Wei C, Macher-Goeppinger S, Schwenger V, Zeier M, Reiser J: Management of severe recurrent focal segmental glomerulosclerosis through circulating soluble urokinase receptor modification. Am J Ther 2013; 20:226–229.
22 Alachkar N, Wei C, Arend LJ, Jackson AM, Racusen LC, Fornoni A, Burke G, Rabb H, Kakkad K, Reiser J, Estrella MM: Podocyte effacement closely links to suPAR levels at time of posttransplantation focal segmental glomerulosclerosis occurrence and improves with therapy. Transplantation 2013;96:649–656.
23 Sever S, Trachtman H, Wei C, Reiser J: Is there clinical value in measuring suPAR levels in FSGS? Clin J Am Soc Nephrol 2013;8: 1273–1275.
24 Maas RJ, Wetzels JF, Deegens JK: Serum-soluble urokinase receptor concentration in primary FSGS. Kidney Int 2012;81:1043–1044.
25 Bock ME, Price HE, Gallon L, Langman CB: Serum-soluble urokinase-type plasminogen activator receptor levels and idiopathic FSGS in children: a single-center report. Clin J Am Soc Nephrol 2013;8:1304–1311.
26 Naesens M, Meijers B, Sprangers B: suPAR and FSGS: the gap between bench and bedside. Transplantation 2013;96:368–369.
27 Sinha A, Bajpai J, Saini S, Bhatia D, Gupta A, Puraswani M, Dinda AK, Agarwal SK, Sopory S, Pandey RM, Hari P, Bagga A: Serum-soluble urokinase receptor levels do not distinguish focal segmental glomerulosclerosis from other causes of nephrotic syndrome in children. Kidney Int 2014, Epub ahead of print.
28 Wada T, Nangaku M, Maruyama S, Imai E, Shoji K, Kato S, Endo T, Muso E, Kamata K, Yokoyama H, Fujimoto K, Obata Y, Nishino T, Kato H, Uchida S, Sasatomi Y, Saito T, Matsuo S: A multicenter cross-sectional study of circulating soluble urokinase receptor in Japanese patients with glomerular disease. Kidney Int 2014, Epub ahead of print.
29 Meijers B, Maas RJ, Sprangers B, Claes K, Poesen R, Bammens B, Naesens M, Deegens JK, Dietrich R, Storr M, Wetzels JF, Evenepoel P, Kuypers D: The soluble urokinase receptor is not a clinical marker for focal segmental glomerulosclerosis. Kidney Int 2014, Epub ahead of print.
30 Segarra A, Jatem E, Quiles MT, Arbos MA, Ostos H, Valtierra N, Carnicer C, Agraz I, Salcedo MT: Diagnostic value of soluble urokinase-type plasminogen activator receptor serum levels in adults with idiopathic nephrotic syndrome. Nefrologia 2014;34:46–52.
31 Jefferson JA, Shankland SJ: Has the circulating permeability factor in primary FSGS been found? Kidney Int 2013;84:235–238.

Prof. Dr. Jochen Reiser
Department of Internal Medicine
Rush University
Chicago, IL 60612 (USA)
E-Mail Jochen_Reiser@rush.edu

Liu Z-H, He JC (eds): Podocytopathy. Contrib Nephrol. Basel, Karger, 2014, vol 183, pp 132–144
DOI: 10.1159/000360049

Anti-Phospholipase A_2 Receptor-1 Autoantibodies and Membranous Nephropathy

Laurence H. Beck

Boston University School of Medicine, Boston, Mass., USA

Abstract

Background: Membranous nephropathy (MN) is a common cause of adult nephrotic syndrome that is caused by immune-mediated injury to the podocyte. This review highlights the recent advances in our understanding of the primary form of MN, now known to be an organ-specific autoimmune disease. ***Summary:*** The M-type phospholipase A_2 receptor-1 (PLA_2R) has been identified as the major target antigen in primary MN. Approximately 70–80% of patients with clinically active MN have circulating anti-PLA_2R autoantibodies, as detected by Western blot, ELISA, or immunofluorescence assay. The PLA_2R antigen itself co-localizes with IgG in the immune deposits of patients with PLA_2R-associated disease, even after the disappearance of circulating autoantibodies, and such staining in a peripheral capillary loop pattern on kidney biopsy is also indicative of PLA_2R-associated MN. Genetic studies have identified disease risk alleles in the *PLA2R1* gene, as well as in *HLA-DQA1*. Although no specific causative mutation has been identified in *PLA2R1*, there appears to be an interaction between the two gene loci, which may confer susceptibility to the disease, perhaps in the setting of an additional environmental factor. There is much circumstantial evidence implicating anti-PLA_2R antibodies in the pathogenesis of primary MN, but direct evidence is still lacking. ***Key Messages:*** The identification of PLA_2R as the target antigen in primary MN has had major implications for the diagnosis and monitoring of PLA_2R-associated MN, the most common form of primary MN. Further research into this antigen-autoantibody system should provide a better understanding of the pathogenesis, classification, and treatment of primary MN.

Membranous nephropathy (MN) is a frequent cause of adult nephrotic syndrome. In developed countries, the majority of cases are primary in nature, with the remainder being secondary to systemic autoimmune diseases such as lupus, infections such as hepatitis B, a limited number of therapeutic agents (nonsteroidal anti-inflammatory agents being the most common), and malignancy. Affected individuals present with the insidious onset of edema, proteinuria, hypoalbuminemia, and hyperlipidemia, typically after the disease process has been ongoing for at least several months. More infrequently, patients will acutely present to clinical attention as a result of a deep venous thrombosis or pulmonary embolism, which are particularly common in MN. Although one third of cases may ultimately undergo spontaneous remission, many others require treatment with immunosuppressive agents due to severe proteinuria or worsening renal function. End-stage renal disease may occur as a result of progressive MN, and it is important to recognize that the disease can recur in the kidney allograft. Because of this variable natural history, uncertainty as to who will develop worsening kidney disease, and the risks of systemic immunosuppression, individualized decisions about immunosuppressive treatment are often challenging.

Pathology of Membranous Nephropathy

The podocyte is the main target of disease activity in all forms of MN, and injury is primarily mediated by antibodies and one or more arms of the complement system. The histopathological term 'membranous nephropathy' describes the thickened glomerular basement membranes (GBM) visible by light microscopy. This expansion of the GBM is due to the presence of two abnormal constituents: immunoglobulin-containing immune deposits which initially localize in a subepithelial position at the basal surface of the podocyte, and increased matrix material secondarily secreted by the injured podocyte. Differential staining of these components by the Jones' methenamine silver stain creates a 'spike and crater' pattern due to the silver-avid matrix components and the negatively-staining immune deposits. Immunofluorescence for immunoglobulin or complement factor 3 (C3) results in the peripheral capillary loop granular staining pattern typical of MN. Electron microscopy provides the clearest view of these electron-dense deposits and their spatial relationship to the podocyte. Initial deposits form immediately beneath the podocyte (stage I deposits) and the overlying foot processes become effaced due to local cellular injury. With advancing disease, the deposits become separated from each other by intervening matrix (stage II) and later become completely surrounded by matrix (stage III), forming intramembranous deposits. With resolution of disease, the deposits may slowly clear, leaving electron-lucent areas (stage IV).

These deposits were once considered to result exclusively from circulating immune complexes whose size and physicochemical properties allowed them to traverse the GBM and accumulate on the abluminal, subepithelial side of the

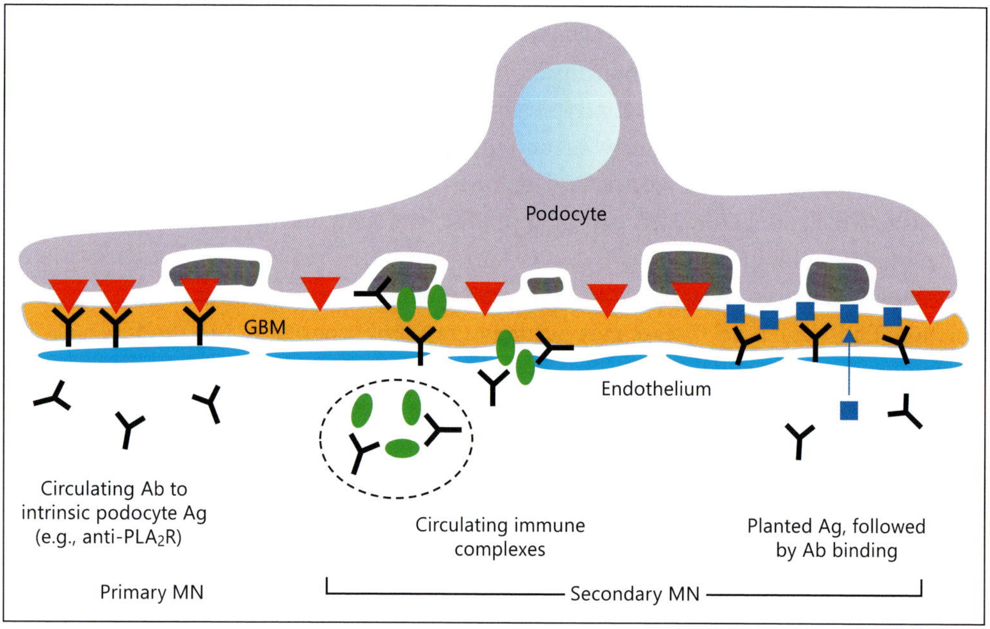

Fig. 1. Potential mechanisms of immune deposit formation in MN. Primary MN is often associated with circulating autoantibodies to an intrinsic podocyte antigen (red triangles) such as PLA_2R. Immune deposits in this situation form in situ due to surface binding of the antibody followed by shedding of antigen-antibody complexes. Secondary disease may involve circulating immune complexes (green ovals) that ultimately deposit in a subepithelial location. Cationic antigens such as cationic bovine serum albumin (see next chapter) may act as planted antigens (blue squares), traversing the GBM and depositing beneath the podocyte by virtue of charge, and then secondarily being targeted by antibodies.

GBM (which may still hold true for lupus-associated MN). However, more recent work in animal models and human disease has suggested that intrinsic podocyte antigens are directly targeted by circulating antibodies to form in situ complexes in many forms of MN (fig. 1). Nevertheless, there is also evidence that proteins extrinsic to the glomerulus can initially localize in a subepithelial position as planted antigens and later be targeted by the humoral immune system (e.g. cationic bovine serum albumin, as described in the following chapter by Ronco and Debiec).

Deposited immune complexes typically activate the classical complement system, whose final common pathways involve the generation of the C5b-9 membrane attack complex (MAC) and elaboration of the anaphylotoxins C3a and C5a. In those glomerular diseases characterized by mesangial or subendothelial deposits, the immune complexes are sufficiently close to the capillary lumen that locally generated C3a and C5a are able to recruit inflammatory cells, resulting in a proliferative glomerulonephritis. In contrast, the subepithelial location of the deposits in MN, combined with the abluminal flow of filtrate, limits the ability of these chemotactic factors to recruit circulating immune cells, and thus MN is generally considered a non-inflammatory disease. For this reason, the often-used term membranous *glomerulonephritis* is a misnomer, and the terms membranous *nephropathy* or *glomerulopathy* are preferred.

Podocyte injury in MN is initiated by the local activation of the complement cascade and insertion of the MAC into the podocyte cell membrane. Local factors such as humoral and membrane-bound complement inhibitors and the ability of the cell to shed areas of membrane damaged by C5b-9 insertion can limit damage, but with increasing mass of immune deposits, the podocyte is overwhelmed and a number of cytopathic changes occur. Calcium influx through the MAC stimulates intracellular signaling pathways that lead to a more dedifferentiated cell phenotype, loss of foot process and slit-diaphragm structures, and elaboration of a more immature basement membrane material [1]. Unlike minimal change disease, in which podocyte phenotype can rapidly revert to normal with resolution of the disease process, the restoration of a normal glomerular architecture is slower in MN due to the structural changes in the GBM, such as the immune deposits, that may take months to years to disappear. Thus, clinical signs such as proteinuria may persist well after a remission of the underlying immune process in MN.

Much of this pathophysiological paradigm was first established in an experimental rat model of MN known as Heymann nephritis [reviewed in 2]. It was ultimately discovered that the target antigen in this model was a large transmembrane receptor known as megalin, and that antibodies directed against this protein formed in situ immune deposits and activated the classical complement pathway. Megalin, however, is not expressed by human podocytes, and it was several decades before the first relevant podocyte antigen, neutral endopeptidase, was identified in human disease (see the following chapter by Ronco and Debiec). It was not until recently that the major antigen in human primary MN was identified as the M-type phospholipase A_2 receptor-1 (PLA_2R), establishing MN as an organ-specific autoimmune disease [3].

Identification of PLA_2R as the Target Antigen in Primary Membranous Nephropathy

One of the reasons that the major MN antigen had remained elusive was that the epitopes in PLA_2R recognized by human autoantibodies depend on a conformation stabilized by intra-chain disulfide bonds. A common protocol for gel electrophoresis includes the addition of reducing agents during sample preparation, and it is conceivable that previous attempts to detect this antigen failed due to loss of the epitopes. When we performed electrophoresis and Western blotting under non-reducing conditions, a 180-kDa band was consistently detected that appeared to be specific for MN [3]. This band, after partial purification and mass spectrometric analysis, was ultimately identified as PLA_2R, a large transmembrane glycoprotein and putative receptor for small, secreted phospholipase A_2 molecules. Earlier cloning and expression analysis had shown that PLA_2R was highly expressed in kidney cortex [4]. Once identified as a potential target antigen, its expression in the glomerulus was shown to be limited to the cell body and processes of the human podocyte [3]. The function of PLA_2R in the podocyte has not yet been established.

PLA_2R is a member of the mannose receptor family and is composed of an N-terminal cysteine-rich domain, a fibronectin-like II domain, eight C-type lectin-like domains, a transmembrane region, and a short cytoplasmic tail containing motifs used in endocytic recycling. It was cloned based on its ability to bind secreted phospholipase A_2 molecules, but its exact function is not fully known. An additional role for PLA_2R in promoting replicative senescence has been identified in human dermal fibroblasts [5]. This finding has recently been extended into the cancer field with the suggestion that PLA_2R may act as a tumor suppressor. Breast cancer cell lines have decreased expression of PLA_2R compared to primary mammary epithelium and forced PLA_2R expression in these cell lines leads to apoptosis [6]. PLA_2R can inhibit cellular transformation with the Ras oncogene via a mechanism involving the Jak2/Stat pathway [7]. The cytoplasmic tail of PLA_2R is not needed for this effect, suggesting the existence of a binding partner. This possibility is strengthened by the knowledge that another mannose receptor family member, MRC2 (mannose receptor C, type 2 or Endo180) complexes with the urokinase receptor (uPAR) and urokinase [8]. It is tempting to speculate that PLA_2R may play a role in the post-mitotic, differentiated state of the podocyte.

Anti-PLA_2R Autoantibodies

The case for PLA_2R being the main target antigen in primary MN is strong, although direct pathogenicity of anti-PLA_2R antibodies still remains to be established in animal or in vitro models. As many as 70–80% of patients with primary disease have circulating anti-PLA_2R antibodies, whose presence is related to clinical disease activity, being high in the nephrotic state, absent in remission, and present during relapses in the native kidney or recurrences in the kidney allograft

[9, 10]. Changes in antibody levels precede clinical changes in proteinuria [3, 11], which may argue for a direct role in pathogenicity. A number of recent articles have reported the prevalence of anti-PLA_2R antibodies in primary MN cases and have detailed the associations between titers of anti-PLA_2R and disease outcome or the chance of attaining a spontaneous remission [3, 9, 11–19]. Interestingly, there do not appear to be major ethnic differences to the prevalence of these antibodies; similar percentages have been found in US and European Caucasians, African-Americans, Chinese, and Koreans. In cross-sectional primary MN cohorts, unselected for disease activity, anti-PLA_2R tends to be present in 45–55% of cases. In more selected populations, such as incident cases assayed at the time of biopsy, or in those with nephrotic range proteinuria, the prevalence increases to 70–80% [11, 13, 17–19]. In addition to its serological presence, anti-PLA_2R has been eluted from biopsy tissues of primary MN patients [3]. It is very unlikely that such autoantibodies are induced by podocyte injury per se, since circulating or tissue-bound anti-PLA_2R antibodies are not found in other forms of MN or other glomerular diseases marked by podocyte injury.

Tissue Staining for PLA_2R in Membranous Nephropathy

Another piece of evidence implicating antibodies to PLA_2R in the disease process is the presence of the antigen within immune deposits (fig. 2). This was initially shown in our initial description of anti-PLA_2R autoantibodies, in which the PLA_2R antigen was shown to co-localize with human IgG in the deposits of primary, but not secondary, MN [3]. These findings were extended with the demonstration that this granular pattern of tissue PLA_2R staining could also be present in biopsies of patients who were seronegative for anti-PLA_2R at the time of sampling [20] and could serve as a historical marker of PLA_2R-associated disease, even after the disease had gone into an immune remission. Due to the availability of a commercial antibody to PLA_2R, this technique has become more commonplace and is increasingly being used as a marker of primary, PLA_2R-associated disease by renal pathologists [21].

This pattern of staining has been called 'enhanced' PLA_2R staining [15] because the majority of detectable PLA_2R is found within the GBM deposits rather than in the podocyte cell body. It is not clear whether the relative amounts of cellular versus deposited antigen have changed, or whether the partially denatured, concentrated antigen is merely a better substrate for the binding of the commercial, polyclonal antibody. Transcript levels of PLA_2R1, as measured by PCR from biopsy tissue, do not appear to be different between cases of MN with or without this 'enhanced' PLA_2R tissue staining [15]. There is also no apparent

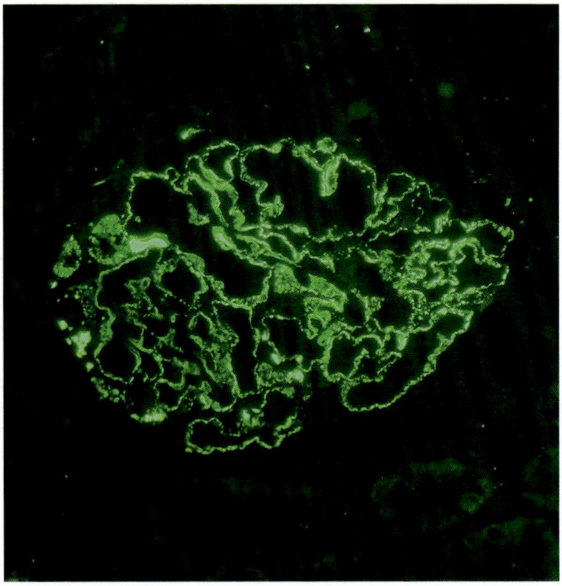

Fig. 2. Immunofluorescence for PLA2R in a biopsy specimen of a case of primary MN. Staining with a commercial antibody to PLA_2R (Sigma Life Science, St. Louis, Mo., USA) reveals bright staining of the antigen within immune deposits, in a peripheral capillary loop pattern. ×400. Image courtesy of Dr. Rivka Ayalon, Renal Section, Boston University School of Medicine.

steric inhibition of the binding of the commercial antibody by anti-PLA_2R autoantibodies that may have engaged PLA_2R on the podocyte membrane [15].

The specificities of anti-PLA_2R seropositivity and the 'enhanced' PLA_2R tissue staining for primary MN have been questioned by their infrequent association with MN cases felt to be secondary in nature. Although some articles [3, 11] report uniform seronegativity for anti-PLA_2R in secondary MN (as well as in other glomerular diseases and healthy controls), others describe several cases of suspected secondary MN associated with anti-PLA_2R antibodies [13, 19] or positive PLA_2R tissue staining [21]. In these rare cases of apparent PLA_2R-associated secondary disease, there is usually an IgG4 predominance (or co-dominance) to the deposits and an absence of C1q staining. It is very likely that, in these cases, the pathology is more reflective of primary MN than the secondary association, which may rather represent coincidental disease. While MN associated with lupus or hepatitis B is usually non-PLA_2R-associated and therefore truly secondary in nature, the reported secondary associations with hepatitis C and sarcoidosis may turn out to be coincidental, as the majority of these cases are PLA_2R-associated [21]. To help distinguish whether a case is primary or secondary, features such as the presence of circulating anti-PLA_2R, or the detection of

PLA_2R and IgG4 in immune deposits, can suggest primary disease, whereas the presence of subendothelial deposits or a full-house immunofluorescence staining pattern would suggest secondary MN [21].

Genetic Associations

A final piece of evidence linking PLA_2R to primary MN is the presence of genetic risk alleles in *PLA2R1*, the gene coding for the antigen. Two small Asian studies used a targeted genotyping approach to reveal an enrichment of several single nucleotide polymorphisms within the *PLA2R1* gene in cases of MN versus controls [22, 23]. Subsequently, an unbiased genome-wide association study using three cohorts of European Caucasians revealed major risk loci in the *HLA-DQA1* and *PLA2R1* genes [24]. Sequencing of all exons in the *PLA2R1* gene failed to detect a causative allele that might affect the protein configuration [25]. Genotype-phenotype correlations have shown that those who carry a high-risk haplotype in the *HLA-DQA1* and *PLA2R1* genes have a much higher prevalence of circulating anti-PLA_2R [26]. Another study has also demonstrated higher anti-PLA_2R titers in subjects carrying one or two risk alleles for HLA DQA1*05:01 or for DQB1*02:01 than in those with no risk alleles [17]. It is interesting that the 'risk' alleles at the *PLA2R1* locus are actually the major alleles, indicating that a large proportion of the population might be at risk for the disease. Some have proposed that a confluence of factors, including genetic risk alleles, environmental factors, and perhaps age-related changes in immunoglobulin glycosylation, are all necessary for the development of disease [27].

Unique Pathological Features

It is generally assumed, due to the presence of C3, C4, and C5b-9 within immune deposits as well as decades of study in the Heymann nephritis model [1, 2], that complement activation represents a major mechanism of podocyte injury in human MN. The classical pathway, however, may not be of primary importance, since C1q staining is typically absent to weak in primary and/or PLA_2R-associated MN [21]. With the exception of very early disease, in which case IgG1 may predominate [28], IgG4 is the major IgG found in immune deposits and is the major circulating form of anti-PLA_2R [3, 16, 17]. IgG4 represents a peculiar subclass of IgG as it can form an asymmetric molecule with two distinct antigenic specificities through Fab arm exchange [29]; this property is felt to be anti-inflammatory by limiting the size of immune complexes. IgG4 also does not appear to activate the classical complement system. Our laboratory has shown preliminary evidence that IgG4 may instead directly activate the lectin pathway of complement through the binding mannan-binding lectin to IgG4 molecules with hypogalactosylated glycan moieties [30]. An exceptional case of MN caused by monoclonal IgG3 κ anti-PLA_2R reveals that activation of complement via the classical pathway (demonstrated by the presence of C1q in deposits) can phenocopy the more typical IgG4-predominant primary MN [31]. There are also cases of (polyclonal) PLA_2R-associated MN in which no circulating IgG4 anti-PLA_2R is found [16], or that lack IgG4 in the deposits [15]. The precise mechanisms of complement activation remain an area of active investigation.

Natural History and Treatment of PLA_2R-Associated Membranous Nephropathy

The mean age of disease onset for PLA_2R-associated MN is in the early 50s [14–17, 20]. However, PLA_2R-associated disease has been reported in adolescents [11, 14] and children as young as 10 years of age [32]. As mentioned previously, it can occur in all races and ethnicities. One third of the cases will undergo spontaneous remission, and measuring anti-PLA_2R titers may be informative as to the likelihood of spontaneous remission [16]. For enduring disease, especially with worsening renal function and/or complications of the nephrotic syndrome, treatment with immunosuppressive agents is often indicated. Alkylating agents such as cyclophosphamide in combination with steroids, or calcineurin inhibitors remain the mainstay of treatment [33]. The anti-B-cell agent rituximab, despite the lack of randomized, controlled trials, also appears to be a promising treatment for this disease.

Treatment decisions are often difficult in MN due to the limited ability to predict whether a patient will spontaneously remit, or progress to worsened renal function, and over what time course. Most immunosuppressive regimens have major adverse effects, and these risks must be balanced against the benefits of treatment. Knowledge of anti-PLA_2R titers, especially when measured longitudinally, may be informative as to whether immunosuppressive treatment should be pursued (table 1). If a patient is known to have PLA_2R-associated disease by virtue of tissue PLA_2R staining of the biopsy, but has negative or declining PLA_2R titers, then spontaneous remission has occurred or is in progress. A patient with stable or progressively increasing moderate-to-high titers of anti-PLA_2R is unlikely to remit, and treatment decisions should be made according to clinical guidelines [33]; concerning features include high-degree proteinuria,

Table 1. Suggested use of anti-PLA_2R serology and PLA_2R staining of biopsy specimens to define MN type and treatment options. In those patients who are seropositive for anti-PLA_2R, measurement of serial titers can indicate whether or not the disease is entering immunologic remission (either spontaneously or as a result of prior therapy). High titers of anti-PLA_2R have been associated with a lower likelihood of spontaneous remission [16]

Tissue PLA_2R[a]	Serum anti-PLA_2R	Interpretation	Disease activity	Therapeutic action
POS	Neg	PLA_2R-associated MN	In immunologic remission	Supportive therapy[b]
POS	POS (declining titers)	PLA_2R-associated MN	Entering immunologic remission	Supportive therapy
POS	POS (high or rising titers)	PLA_2R-associated MN	Active disease	Consider immunosuppressive therapy, according to clinical guidelines[c]
Neg	Neg	Possible idiopathic MN[d]	Determined by clinical features	Follow clinical guidelines, including possibility of immunosuppression
Neg	Neg	Possible secondary MN[e]	Determined by clinical features	Treat secondary cause, if known

[a] Tissue PLA_2R indicates the presence of PLA_2R, as detected by immunofluorescence or immunohistochemistry of a biopsy specimen, in a granular, capillary loop pattern.
[b] Supportive therapy includes the use of inhibitors of the renin-angiotensin-aldosterone system, diuretics, cholesterol-lowering agents, and close monitoring of clinical status.
[c] Guidelines for treatment of MN can be found in the 2012 KDIGO Clinical Practice Guidelines for Glomerulonephritis.
[d] The presence of IgG4 predominance in immune deposits, and the absence of secondary features or associations would suggest idiopathic MN.
[e] Secondary associations include lupus, hepatitis B, malignancy, among others. The presence of subendothelial deposits, a full-house immunofluorescence staining pattern, and the dominance of an IgG subclass other than IgG4 would also be suggestive of secondary MN.

worsening renal function, or major complications of the nephrotic syndrome. A patient with MN who is both seronegative for anti-PLA_2R and tissue-negative for PLA_2R in immune deposits may have truly idiopathic primary MN (e.g. if the predominant IgG subclass is IgG4) or may have a secondary cause of MN. Prior to initiating immunosuppressive treatment, reasonable efforts should be made to exclude secondary disease. Finally, MN can recur in the kidney allograft and circulating anti-PLA_2R or tissue staining for PLA_2R is often positive [10, 34]. Standard transplant immunosuppression may limit the development or severity of recurrence, but it is clear that in certain cases, a full-blown nephrotic syndrome from recurrent MN can take place. Case reports have suggested a benefit of rituximab in these situations.

Conclusion

The identification of PLA_2R as the major target antigen in primary MN has ushered in a new era of investigation into the diagnosis, monitoring, treatment and pathophysiology of primary MN. We look forward to better understanding of the role of this molecule not only in disease, but also in the normal physiology of the podocyte.

Acknowledgment

L.H.B.'s research is supported by grant support from the NIH-NIDDK (R01 DK097053).

Disclosure Statement

The author is a co-inventor of US patent No. 8,507,215 'Diagnostics for membranous nephropathy'.

References

1 Cybulsky AV: Membranous nephropathy. Contrib Nephrol. Basel, Karger, 2011, vol 169, pp 107–125.

2 Jefferson JA, Pippin JW, Shankland SJ: Experimental models of membranous nephropathy. Drug Discov Today Dis Models 2010;7: 27–33.

3 Beck LH Jr, Bonegio RG, Lambeau G, Beck DM, Powell DW, Cummins TD, Klein JB, Salant DJ: M-type phospholipase A_2 receptor as target antigen in idiopathic membranous nephropathy. N Engl J Med 2009;361:11–21.

4 Ancian P, Lambeau G, Mattei MG, Lazdunski M: The human 180-kDa receptor for secretory phospholipases A_2. Molecular cloning, identification of a secreted soluble form, expression, and chromosomal localization. J Biol Chem 1995;270:8963–8970.

5 Augert A, Payre C, de Launoit Y, Gil J, Lambeau G, Bernard D: The M-type receptor PLA_2R regulates senescence through the p53 pathway. EMBO Rep 2009;10:271–277.

6 Augert A, Vindrieux D, Girard CA, Le Calvé B, Gras B, Ferrand M, Bouchet BP, Puisieux A, de Launoit Y, Simonnet H, Lambeau G, Bernard D: PLA_2R1 kills cancer cells by inducing mitochondrial stress. Free Radic Biol Med 2013;65C:969–977.

7 Vindrieux D, Augert A, Girard CA, Gitenay D, Lallet-Daher H, Wiel C, Le Calve B, Gras B, Ferrand M, Verbeke S, de Launoit Y, Leroy X, Puisieux A, Aubert S, Perrais M, Gelb M, Simonnet H, Lambeau G, Bernard D: PLA_2R1 mediates tumor suppression by activating JAK2. Cancer Res 2013;73:6334–6345.

8 Behrendt N: The urokinase receptor (uPAR) and the uPAR-associated protein (uPARAP/Endo180): membrane proteins engaged in matrix turnover during tissue remodeling. Biol Chem 2004;385:103–136.

9 Hofstra JM, Beck LH Jr, Beck DM, Wetzels JF, Salant DJ: Anti-phospholipase A_2 receptor antibodies correlate with clinical status in idiopathic membranous nephropathy. Clin J Am Soc Nephrol 2011;6:1286–1291.

10 Debiec H, Martin L, Jouanneau C, Dautin G, Mesnard L, Rondeau E, Mousson C, Ronco P: Autoantibodies specific for the phospholipase A_2 receptor in recurrent and de novo membranous nephropathy. Am J Transplant 2011; 11:2144–2152.
11 Hoxha E, Harendza S, Zahner G, Panzer U, Steinmetz O, Fechner K, Helmchen U, Stahl RA: An immunofluorescence test for phospholipase-A_2 receptor antibodies and its clinical usefulness in patients with membranous glomerulonephritis. Nephrol Dial Transplant 2011;26:2526–2532.
12 Beck LH Jr, Fervenza FC, Beck DM, Bonegio RG, Malik FA, Erickson SB, Cosio FG, Cattran DC, Salant DJ: Rituximab-induced depletion of anti-PLA_2R autoantibodies predicts response in membranous nephropathy. J Am Soc Nephrol 2011;22:1543–1550.
13 Qin W, Beck LH Jr, Zeng C, Chen Z, Li S, Zuo K, Salant DJ, Liu Z: Anti-phospholipase A_2 receptor antibody in membranous nephropathy. J Am Soc Nephrol 2011;22:1137–1143.
14 Dähnrich C, Komorowski L, Probst C, Seitz-Polski B, Esnault V, Wetzels JF, Hofstra JM, Hoxha E, Stahl RA, Lambeau G, Stöcker W, Schlumberger W: Development of a standardized ELISA for the determination of autoantibodies against human M-type phospholipase A_2 receptor in primary membranous nephropathy. Clin Chim Acta 2013;421C:213–218.
15 Hoxha E, Kneissler U, Stege G, Zahner G, Thiele I, Panzer U, Harendza S, Helmchen UM, Stahl RA: Enhanced expression of the M-type phospholipase A_2 receptor in glomeruli correlates with serum receptor antibodies in primary membranous nephropathy. Kidney Int 2012;82:797–804.
16 Hofstra JM, Debiec H, Short CD, Pellé T, Kleta R, Mathieson PW, Ronco P, Brenchley PE, Wetzels JF: Antiphospholipase A_2 receptor antibody titer and subclass in idiopathic membranous nephropathy. J Am Soc Nephrol 2012;23:1735–1743.
17 Kanigicherla D, Gummadova J, McKenzie EA, Roberts SA, Harris S, Nikam M, Poulton K, McWilliam L, Short CD, Venning M, Brenchley PE: Anti-PLA_2R antibodies measured by ELISA predict long-term outcome in a prevalent population of patients with idiopathic membranous nephropathy. Kidney Int 2013;83:940–948.
18 Svobodova B, Honsova E, Ronco P, Tesar V, Debiec H: Kidney biopsy is a sensitive tool for retrospective diagnosis of PLA_2R-related membranous nephropathy. Nephrol Dial Transplant 2012;28:1839–1844.
19 Oh YJ, Yang SH, Kim DK, Kang SW, Kim YS: Autoantibodies against phospholipase A_2 receptor in Korean patients with membranous nephropathy. PLoS One 2013;8:e62151.
20 Debiec H, Ronco P: PLA_2R autoantibodies and PLA_2R glomerular deposits in membranous nephropathy. N Engl J Med 2011;364: 689–690.
21 Larsen CP, Messias NC, Silva FG, Messias E, Walker PD: Determination of primary versus secondary membranous glomerulopathy utilizing phospholipase A_2 receptor staining in renal biopsies. Mod Pathol 2012;26:709–715.
22 Kim S, Chin HJ, Na KY, Oh J, Chung W, Noh JW, Lee YK, Cho JT, Lee EK, Chae DW: Single nucleotide polymorphisms in the phospholipase A_2 receptor gene are associated with genetic susceptibility to idiopathic membranous nephropathy. Nephron Clin Pract 2010;117:c253–c258.
23 Liu YH, Chen CH, Chen SY, Lin YJ, Liao WL, Tsai CH, Wan L, Tsai FJ: Association of phospholipase A_2 receptor-1 polymorphisms with idiopathic membranous nephropathy in Chinese patients in Taiwan. J Biomed Sci 2010;17:81.
24 Stanescu HC, Arcos-Burgos M, Medlar A, Bockenhauer D, Kottgen A, Dragomirescu L, Voinescu C, Patel N, Pearce K, Hubank M, Stephens HA, Laundy V, Padmanabhan S, Zawadzka A, Hofstra JM, Coenen MJ, den Heijer M, Kiemeney LA, Bacq-Daian D, Stengel B, Powis SH, Brenchley P, Feehally J, Rees AJ, Debiec H, Wetzels JF, Ronco P, Mathieson PW, Kleta R: Risk HLA-DQA1 and PLA_2R1 alleles in idiopathic membranous nephropathy. N Engl J Med 2011;364:616–626.
25 Coenen MJ, Hofstra JM, Debiec H, Stanescu HC, Medlar AJ, Stengel B, Boland-Augé A, Groothuismink JM, Bockenhauer D, Powis SH, Mathieson PW, Brenchley PE, Kleta R, Wetzels JF, Ronco P: Phospholipase A_2 receptor (PLA_2R1) sequence variants in idiopathic membranous nephropathy. J Am Soc Nephrol 2013;24:677–683.

26 Lv J, Hou W, Zhou X, Liu G, Zhou F, Zhao N, Hou P, Zhao M, Zhang H: Interaction between PLA_2R1 and HLA-DQA1 variants associates with anti-PLA_2R antibodies and membranous nephropathy. J Am Soc Nephrol 2013;24:1323–1329.
27 Salant DJ: Genetic variants in membranous nephropathy: perhaps a perfect storm rather than a straightforward conformeropathy? J Am Soc Nephrol 2013;24:525–528.
28 Huang CC, Lehman A, Albawardi A, Satoskar A, Brodsky S, Nadasdy G, Hebert L, Rovin B, Nadasdy T: IgG subclass staining in renal biopsies with membranous glomerulonephritis indicates subclass switch during disease progression. Mod Pathol 2013;26: 799–805.
29 Van der Neut Kolfschoten M, Schuurman J, Losen M, Bleeker WK, Martínez-Martínez P, Vermeulen E, den Bleker TH, Wiegman L, Vink T, Aarden LA, De Baets MH, van de Winkel JG, Aalberse RC, Parren PW: Anti-inflammatory activity of human IgG4 antibodies by dynamic Fab arm exchange. Science 2007;317:1554–1557.
30 Ma H, Beck LH, Salant DJ: Membranous nephropathy-associated anti-phospholipase A_2 receptor IgG4 autoantibodies activate the lectin complement pathway (abstract). J Am Soc Nephrol 2011;22:62A.
31 Debiec H, Hanoy M, Francois A, Guerrot D, Ferlicot S, Johanet C, Aucouturier P, Godin M, Ronco P: Recurrent membranous nephropathy in an allograft caused by IgG3κ targeting the PLA_2 receptor. J Am Soc Nephrol 2012;23:1949–1954.
32 Cossey LN, Walker PD, Larsen CP: Phospholipase A_2 receptor staining in pediatric idiopathic membranous glomerulopathy. Pediatr Nephrol 2013;28:2307–2311.
33 Kidney Disease: Improving Global Outcomes (KDIGO) Glomerulonephritis Work Group: KDIGO clinical practice guideline for glomerulonephritis. Kidney Int Suppl 2012;2: 139–274.
34 Larsen CP, Walker PD: Phospholipase A_2 receptor (PLA_2R) staining is useful in the determination of de novo versus recurrent membranous glomerulopathy. Transplantation 2013;95:1259–1262.

Laurence H. Beck, Jr, MD, PhD
Renal Section, 5th floor, X-504
650 Albany Street
Boston, MA 02118 (USA)
E-Mail lhbeckjr@bu.edu

Liu Z-H, He JC (eds): Podocytopathy. Contrib Nephrol. Basel, Karger, 2014, vol 183, pp 145–156
DOI: 10.1159/000360051

Anti-Neutral Endopeptidase Antibodies, Autoantibodies, Anti-Bovine Serum Albumin Antibodies and Membranous Nephropathy

P. Ronco[a–c] • H. Debiec[a–c]

[a]INSERM UMR S 1155, [b]Sorbonne Universités, UPMC Univ Paris 06, and [c]AP-HP, Hôpital Tenon, Service de Néphrologie et Dialyses, Paris, France

Abstract

Membranous nephropathy (MN) is characterized by an accumulation of immune deposits on the subepithelial side of the glomerular basement membrane, which results in complement activation and proteinuria. Since 2002, several major antigens of the podocyte have been identified in human MN. The first antigen is neutral endopeptidase (NEP), the alloantigen involved in neonatal cases of MN that occur in newborns from NEP-deficient mothers. The second antigen, which is also the first autoantigen identified in idiopathic MN in the adult, is the M-type phospholipase A_2 receptor (PLA_2R). Both NEP and PLA_2R are present on the podocyte surface where they can serve as targets for circulating antibodies. Antibodies against cytoplasmic podocytic antigens, such as aldose reductase, superoxide dismutase-2 and α-enolase, have also been reported although their role in the pathogenesis remains controversial because they are not present at the podocyte membrane in normal conditions. In addition to podocyte antigens, a circulating antigen, cationic bovine serum albumin, was shown to be implicated in cases of early childhood MN. This suggests that dietary antigens may be involved through charge-dependent binding to anionic glomerular capillary wall and in situ formation of immune complexes. This review will be focused on non-PLA_2R antigens.

Membranous nephropathy (MN) is characterized by an accumulation of immune deposits on the outer aspect of the glomerular capillary wall. The immune deposits consist of IgG, mainly IgG4 and IgG1 [1], of antigens that have long eluded identification and of the membrane attack complex of complement C5b-9. The formation of subepithelial immune deposits results in complement activation which is responsible for functional impairment of the glomerular capillary wall causing proteinuria [2]. The so-called 'idiopathic' MN, without any identified etiology, is an autoimmune disease, while secondary forms involve exogenous antigens such as viral and tumoral antigens [3].

Since 2002, major advances have occurred in the understanding of the pathobiology of MN. The first breakthrough was the identification of neutral endopeptidase (NEP) as the responsible antigen in a rare subset of patients with alloimmune antenatal MN [4, 5]. This finding provided the proof of concept that a human podocyte antigen could serve as a target for pathogenic antibodies, as shown in the early 1980s for the rat podocyte megalin in Heymann nephritis, the first experimental model of MN [6]. The identification of the podocyte NEP paved the way for that of M-type phospholipase A_2 receptor (PLA_2R), the first podocyte antigen involved in autoimmune idiopathic MN [7]. More recently, other antigens including aldose reductase and superoxide dismutase-2 (SOD2) have been identified, but their role is controversial because they are not present at the surface of normal podocytes [8]. Non-podocyte circulating antigens can also be involved as is the case for cationic bovine serum albumin (BSA) in children less than 5 years of age [9], who represent the human counterpart to the rabbit model of 'planted' antigen first described by Border et al. [10] in the early 1980s.

This review will focus on NEP, cytoplasmic podocytic antigens and dietary antigens such as BSA (see Beck et al. in this issue for the autoantigen PLA_2R).

Neutral Endopeptidase and Alloimmune Membranous Nephropathy

Pathophysiology

This rare form of MN was first identified in a neonate born with respiratory distress and acute renal failure followed by nephrotic range proteinuria and hypertension [4]. Because of early occurrence of the disease in the last trimester of pregnancy, we hypothesized that the mother became immunized during pregnancy and the pathogenic antibodies were transferred to the fetus. Involvement of NEP antigen was suspected because of interspecies differences in reactivity observed when human, rat and rabbit kidneys were incubated with sera from the mother and her infant. Identification of NEP was confirmed by immunoprecipitation of rat brush-border extracts [4, 5]. Transfer experiments, in which a pregnant rabbit was injected with the mother's IgG fraction, and co-localization of NEP, IgG and C5b-9 within the subepithelial immune deposits, established that the disease was caused by anti-NEP antibodies [4, 11]. NEP is not only an antigen but also an enzyme involved in the degradation of biologically active peptides such as enkephalin (for this reason, it is also called *enkephalinase*), natriuretic peptides, endothelin, bradykinin and substance P in the vicinity of their receptors. In the kidney, NEP is present at the podocyte surface, in the brush border of the proximal tubule and in vessel walls. It is found in numerous organs but in very specific localizations, and in granulocytes as well. NEP is identical to the common acute lymphoblastic leukemia antigen and to CD10.

NEP deficiency was suspected because the child's mother was apparently healthy without renal manifestation despite persistently high titers of anti-NEP antibody. This hypothesis was supported by the lack of reactivity of her granulocyte extracts with a panel of anti-NEP antibodies [4]. We have identified a further four families, all from different countries, with fetomaternal alloimmune glomerulopathy. All immunized mothers studied thus far failed to produce NEP and were devoid of NEP enzymatic activity because of truncating mutations in the *NEP* gene located in exon 7 (detected in all mothers) and in exon 15 (detected only in the first reported mother who was a heterozygous compound) [5]. A founder effect by a common ancestor is suspected because the same mutation in exon 7 was identified in all families. Alloimmunization occurred during previous spontaneous miscarriages or at the time of the ongoing pregnancy during which the mother's immune system was first exposed to the NEP antigen of paternal origin on syncytiotrophoblastic cells and fetal cells. The NEP-deficient individuals had normal blood pressure, renal functional tests, and lymphocyte phenotype and function [5], whereas *NEP* null mice showed hypotension, greater sensitivity to septic shock, early onset Alzheimer's disease and prostatic tumors [12]. Because mutations in the *NEP* gene are asymptomatic, one can expect a high prevalence in selected populations.

After birth, renal failure and the nephrotic syndrome rapidly improved in all children in parallel with the disappearance from the child's blood of maternal IgG. However, a newborn had persisting renal failure requiring dialysis despite exchange transfusions. This case led to conclude that the anti-NEP IgG1 was responsible for the disease because the mother had high titers of anti-NEP IgG1 with very low titer of anti-NEP IgG4 and other IgG subclasses. This finding is in keeping with the capacity of IgG1 to activate complement whereas IgG4 usually fail to do so. Another patient with the longest follow-up developed severe chronic renal failure with nephrotic range proteinuria leading to end-stage renal disease requiring transplantation. Although a repeat kidney biopsy could not be performed, the delayed renal manifestations might be induced by advanced MN combined with the postponed consequences of immunologically mediated antenatal nephron loss. Antibodies against idiotypes or allotypes on the maternal IgG might also have contributed to further progression of the disease. These observations suggest that cases of MN or chronic renal failure diagnosed during

and aldose reductase [8]. The corresponding antigens have been detected in glomerular immune deposits by Ghiggeri's group [8, 16] but they give only a very weak positive signal by immunoelectron microscopy. Whether these antigens are involved in the formation of immune deposits still remains a matter of debate. Our group also found that sera of most patients with MN displayed a wide spectrum of reactivity when tested on subcellular fractions isolated from cultured human podocytes. The immunoreactive spots separated by bi-dimensional electrophoresis were excised and analyzed by LC-MS/MS mass spectrometry. We thus identified proteins involved in metabolic pathways (mainly glycolytic enzymes including α-enolase), cytoskeletal proteins such as actin, tubulin, myosin, and stress induced chaperones. Reactivity of these antigens with sera from patients with MN was confirmed with the respective commercially available recombinant proteins [Debiec and Ronco, pers. commun.]. By using another approach based on immunoscreening of a renal cDNA library, a synaptonemal complex protein, also expressed in the cytoplasm, was identified as a new autoimmune target in idiopathic MN [17]. Studies in the rat model of Heymann nephritis had previously suggested that in addition to megalin, antibody specificities directed against endothelial cells or podocytes might be involved [18]. Likewise in a model of passive Heymann nephritis in mice, many intracellular proteins have been identified as potential antigens [19].

The ubiquitous proteins mentioned above are not, or only very weakly, expressed on the membrane of normal podocytes. Thus, could the antibodies to podocyte cytoplasmic proteins have a pathophysiological relevance? One can hypothesize that following the initial injury induced by complement activation and production of reactive oxygen species, intracellular proteins, cryptic epitopes and neoepitopes may be exposed onto the cell membrane. Such an ectopic expression was shown for SOD2 on podocytes [8], components of the glycolytic pathways on the surface of apoptotic cells [20] and the cytoskeletal protein myosin on muscle cells [21]. Therefore, in disease conditions, antibodies to these cytoplasmic proteins could be pathogenic and might perturb important cellular functions such as glycolysis upon binding to their respective antigen [22]. We recently found that antibodies against a cell stress chaperone acting as a negative regulator of podocyte injury could play an important role in the maintenance and progression of MN [23]. Alternatively, autoantibodies directed against podocyte intracellular proteins may result from epitope spreading with autoantibody diversification as frequently observed in autoimmune diseases. Irrespective of the triggering events, the multiple antigen-antibody systems that are generated during the antibody response might be involved at different time points and in different subcategories of patients with MN (fig. 2).

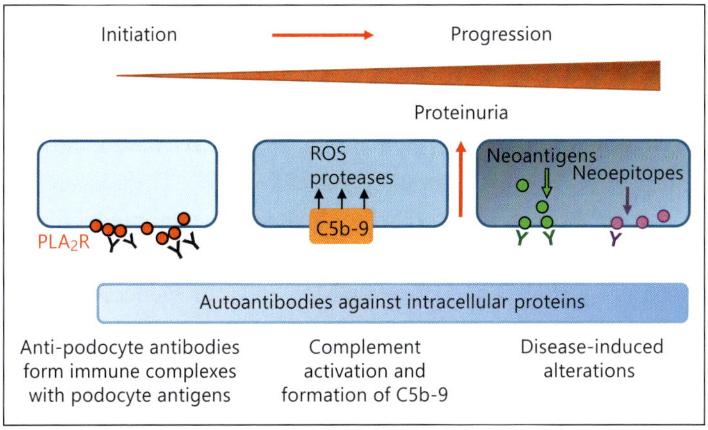

Fig. 2. Mechanisms of immune-mediated podocyte injury and disease progression. Anti-podocyte antibodies targeting integral membrane proteins (e.g. PLA_2R) cross the basement membrane and induce clustering of antigens and shedding of immune complexes in the extracellular space. This process results in complement activation and formation of C5b-9 on the membrane of podocytes, which in turn leads to a variety of intracellular events, including production of reactive oxygen species and proteases, cytoskeletal changes and a stress response. Under disease conditions, cytoplasmic proteins can be routed to the cell membrane and a variety of neoantigens or neoepitopes may be induced by injury. Circulating antibodies raised against those antigens could be nephritogenic and implicated in the perpetuation of the disease.

Another point of particular importance concerns the clinical relevance as biomarkers of antibodies to the podocyte cytoplasmic proteins. In contrast to anti-PLA_2R antibodies, these antibodies are not specific for MN since they are also frequently detected in other autoimmune diseases as well as in healthy individuals [24, 25]. The next step will be to identify the disease-specific epitopes recognized by those anti-cytoplasmic antibodies as has been done in other autoimmune diseases. For example, it was shown that antibodies against the citrullinated form of α-enolase were specific for rheumatoid arthritis [26]. Fine epitope mapping of candidate autoantibodies should be achieved before setting up MN-specific immunoassays and using them routinely in patients with MN.

Until now no comparison of circulating anti-PLA_2R antibodies or deposited PLA_2R antigen with the presence of other antibody and antigen specificities has been carried out in individual patients [27, 28]. Such studies are required to determine whether antibodies to podocyte cytoplasmic proteins will represent an add-on to the assessment of anti-PLA_2R antibodies in blood and of PLA_2R antigen in kidney biopsies and whether they will be useful for patients monitoring.

Bovine Serum Albumin and other Dietary Antigens

MN is one of the most common causes of nephrotic syndrome in adults but this disease is rarely diagnosed in pediatric populations [29, 30]. We have identified BSA as a novel antigen in children with MN aged <5 years [9]. These cases that can be considered as the human counterpart to the experimental model in rats and mice induced by cationized BSA [10, 31, 32], open up new perspectives as to a role of dietary and environmental factors in MN, and more generally in glomerular diseases.

In this model of MN, podocyte antigens do not contribute to the formation of immune deposits which involve an exogenous antigen, BSA, that becomes implanted in the glomerular capillary wall (see Beck et al., this issue). The antigen charge is a key factor for disease induction. Only the animals that were intravenously given chemically modified cationized BSA shortly after immunization with this antigen developed MN [10, 31, 32]. When native, slightly acidic BSA was injected, animals did not develop the disease. Positively charged cationic BSA bound to the anionic glomerular capillary wall because of electrostatic interactions instead of being taken up by endocytosis in podocytes, and could then serve as a target for circulating antibodies.

We speculated that the same pathophysiological scenario might occur in humans. High levels of circulating anti-BSA antibodies of IgG1 and IgG4 subclasses were identified in children and adults with MN whereas anti-BSA IgE was not detected. Circulating BSA antigen was identified in the patients with the higher levels of anti-BSA antibodies. BSA immunopurified from children' sera migrated in the basic range of pH whereas in adult patients, it migrated in the neutral region as native BSA. BSA antigen was co-localized with IgG in subepithelial immune deposits only in the children with both circulating cationic BSA and anti-BSA antibodies, but in none of the adult patients with MN, which strongly supported the assumption that both cationic BSA serving as a 'planted antigen' and concomitant immunization against this antigen could induce MN (fig. 3). Additionally, decrease of cationic BSA and antibody levels during MN remission reinforced a direct causal link with the disease. Contrary to the observation made in the majority of adults with MN, neither PLA_2R autoantibodies nor PLA_2R antigen were detected in the blood and subepithelial immune deposits, respectively, in the children with BSA-related MN [9].

Cow's milk is the major source of BSA in young children. However, the origin of cationic BSA remains obscure. In modern dietary culture, food ingredients are subjected to a variety of processing conditions and contain many additives. Some of these technological processes might induce alterations of food proteins, which could in turn change their physicochemical properties and digestion in

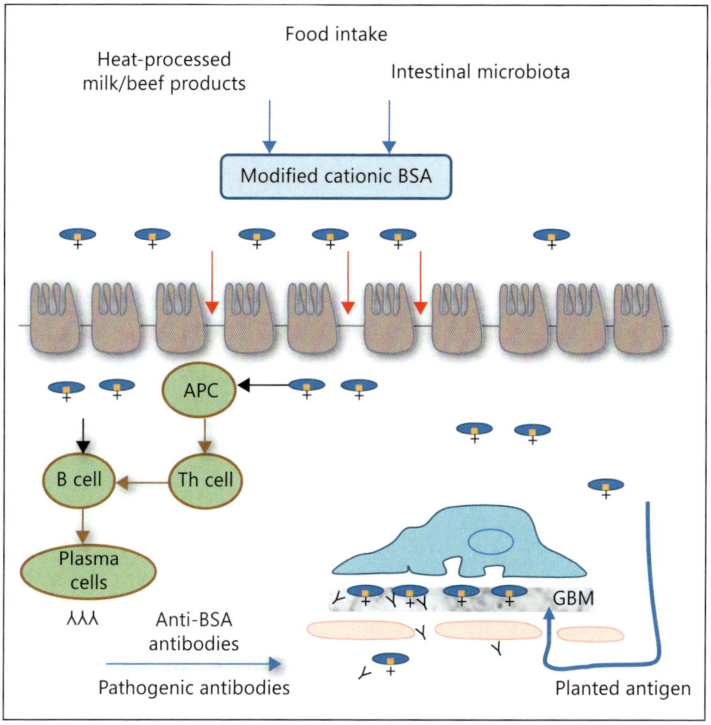

Fig. 3. Proposed pathophysiological mechanisms of cationic BSA-induced MN. MN is initiated by maladaptive immune response against cationic BSA that becomes 'planted' in the subepithelial space because of electrostatic interactions. B cells, with assistance from helper T (Th) cells, differentiate into plasma cells that produce pathogenic antibodies. Upon binding to the planted target antigen, immune complexes are formed on the outer (subepithelial) side of the glomerular basement membrane (GBM).

the gastrointestinal tract [33–35]. On the other hand, it is conceivable that intestinal microbiota and frequent episodes of gastroenteritis as commonly observed in children might lead to pathological modifications of BSA [36].

Although very soluble in water and rich in amino acids recognized by gastrointestinal enzymes, BSA is relatively resistant to digestion. Heat treatment of BSA induces denaturation of the protein and results in reduced proteolysis [37] at the relatively high pH (3–4) of the infants' stomach compared with that of adults (pH 2) [38]. In addition, the amount of intact BSA that enters the intestinal circulation is likely higher during infancy before the gastrointestinal tract has matured and its barrier function has been established [36, 39]. Finally, modification of BSA could directly affect how this protein is cleaved and processed in vivo. Both the predominance of a cationic form and the increased amount of absorbed BSA most likely contribute to the development of MN in young children.

From a clinical point of view, a diagnosis of cationic BSA-related MN should be considered in young children with MN, which should prompt a search for BSA in immune deposits. If BSA is detected, then BSA should first be eliminated from the diet before using more aggressive therapy. The identification of an environmental trigger like food BSA as a causative antigen in young children with MN raises the possibility that other food antigens might be involved.

In conclusion, substantial progresses have recently been made in understanding the pathophysiology of human MN. The first human podocyte antigen, NEP, has been identified in a small subset of patients with neonatal alloimmune disease, followed by the identification of PLA_2R in autoimmune idiopathic MN, and that of BSA in early childhood MN. These major breakthroughs hold the promise of innovative targeted therapies.

Acknowledgments

The research of the authors is supported by grants from Fondation pour la Recherche Médicale, Agence Nationale pour la Recherche (ANR-Programme Blanc SVSE1 (2012) décision n° ANR-12-BSE1-0002-01), AMGEN France, from European Research Council (ERC-2012-ADG_20120314; Grant Agreement No. 322947), and the 7th Framework Programme of the European Community, contract 2012-305608 'European Consortium for High-Throughput Research in Rare Kidney Diseases (EURenOmics)'.

Disclosure Statement

Part of this work was supported by grants from AMGEN France.

References

1 Kuroki A, Shibata T, Honda H, Totsuka D, Kobayashi K, Sugisaki T: Glomerular and serum IgG subclasses in diffuse proliferative lupus nephritis, membranous lupus nephritis, and idiopathic MN. Intern Med 2002;41:936–942.

2 Glassock RJ: The pathogenesis of idiopathic membranous nephropathy: a 50-year odyssey. Am J Kidney Dis 2010;56:157–167.

3 Ponticelli C: Membranous nephropathy. J Nephrol 2007;20:268–287.

4 Debiec H, Guigonis V, Mougenot B, et al: Antenatal membranous glomerulonephritis due to anti-NEP antibodies. N Engl J Med 2002;346:2053–2060.

5 Debiec H, Nauta J, Coulet F, et al: Role of truncating mutations in MME gene in fetomaternal allo-immunization and neonatal glomerulopathies. Lancet 2004;364:1252–1259.

6 Kerjaschki D, Farquhar MG: The pathogenic antigen of Heymann nephritis is a membrane glycoprotein of the renal proximal tubule brush border. Proc Natl Acad Sci USA 1982;79:5557–5561.

7 Beck LH Jr, Bonegio RG, Lambeau G, et al: M-type phospholipase A_2 receptor as target antigen in idiopathic MN. N Engl J Med 2009;361:11–21.

8 Prunotto M, Carnevali ML, Candiano G, et al: Autoimmunity in membranous nephropathy targets aldose reductase and SOD2. J Am Soc Nephrol 2010;21:507–519.
9 Debiec H, Lefeu F, Kemper MJ, Niaudet P, Deschênes G, Remuzzi G, Ulinski T, Ronco P: Early-childhood membranous nephropathy due to cationic bovine serum albumin. N Engl J Med 2011;364:2101–2110.
10 Border WA, Ward HJ, Kamil ES, et al: Induction of membranous nephropathy in rabbits by administration of an exogenous cationic antigen. J Clin Invest 1982;69:451–461.
11 Ronco P, Debiec H: Molecular pathomechanisms of membranous nephropathy: from Heymann nephritis to alloimmunization. J Am Soc Nephrol 2005;16:1205–1213.
12 Lu B, Figini M, Emanueli C, et al: The control of microvascular permeability and blood pressure by neutral endopeptidase. Nat Med 1997;3:904–907.
13 Nortier JL, Debiec H, Tournay Y, et al: Neonatal disease in neutral endopeptidase alloimmunization: lessons for immunological monitoring. Pediatr Nephrol 2006;1:1399–1405.
14 Ronco PM, Ardaillou N, Verroust P, Lelongt B: Pathophysiology of the podocyte: a target and a major player in glomerulonephritis. Adv Nephrol Necker Hosp 1994;23:91–131.
15 Wakui H, Imai H, Komatsuda A, Miura AB: Circulating antibodies against α-enolase in patients with primary membranous nephropathy. Clin Exp Immunol 1999;118:445–450.
16 Bruschi M, Carnevali ML, Murtas C, et al: Direct characterization of target podocyte antigens and autoantibodies in human membranous glomerulonephritis: α-enolase and borderline antigens. J Proteomics 2011;74: 2008–2017.
17 Cavazzini F, Magistroni R, Furci L, et al: Identification and characterization of a new autoimmune protein in membranous nephropathy by immunoscreening of a renal cDNA library. PLoS One 2012;7:e48845.
18 Adler S, Chen X: Anti-Fx1A antibody recognizes a β_1-integrin on glomerular epithelial cells and inhibits adhesion and growth. Am J Physiol 1992;262:F770–F776.
19 Meyer-Schwesinger C, Dehde S, Klug P, et al: Nephrotic syndrome and subepithelial deposits in a mouse model of immune-mediated anti-podocyte glomerulonephritis. J Immunol 2011;187:3218–3229.
20 Ucker DS, Jain MR, Pattabiraman G, Palasiewicz K, Birge RB, Li H: Externalized glycolytic enzymes are novel, conserved, and early biomarkers of apoptosis. J Biol Chem 2012; 287:10325–10343.
21 Zhang M, Alicot EM, Chiu I, Li J, Verna N, Vorup-Jensen T, Kessler B, Shimaoka M, Chan R, Friend D, Mahmood U, Weissleder R, Moore FD, Carroll MC: Identification of the target self-antigens in reperfusion injury. J Exp Med 2006;203:141–152.
22 Magrys A, Anekonda T, Ren G, Adamus G: The role of anti-α-enolase autoantibodies in pathogenicity of autoimmune-mediated retinopathy. J Clin Immunol 2007;27:181–192.
23 Lefeu F, Fasani F, Jouanneau C, Ronco P, Debiec H: Autoantibodies against chaperonin CCT in glomerular immune deposits and sera in patients with membranous nephropathy: possible role in disease progression. Autoantigens Autoantibodies Autoimmunity 2011;7:105–106.
24 Terrier B, Degand N, Guilpain P, Servettaz A, Guillevin L, Mouthon L: Alpha-enolase: a target of antibodies in infectious and autoimmune diseases. Autoimmun Rev 2007;6:176–182.
25 Servettaz A, Guilpain P, Camoin L, et al: Identification of target antigens of antiendothelial cell antibodies in healthy individuals: a proteomic approach. Proteomics 2008;8: 1000–1008.
26 Lundberg K, Venables PJ: Epitope spreading in animal models: array of hope in rheumatoid arthritis and multiple sclerosis. Arthritis Res Ther 2008;10:122.
27 Ronco P, Debiec H, Imai H: Circulating antipodocyte antibodies in membranous nephropathy: pathophysiologic and clinical relevance. Am J Kidney Dis 2013;62:16–19.
28 Murtas C, Bruschi M, Candiano G, et al: Coexistence of different circulating anti-podocyte antibodies in membranous nephropathy. Clin J Am Soc Nephrol 2012;7:1394–1400.
29 Chen A, Frank R, Vento S, et al: Idiopathic membranous nephropathy in pediatric patients: presentation, response to therapy, and long-term outcome. BMC Nephrol 2007;6: 8–11.
30 Avner ED, Harmon WE, Niaudet P: Pediatric Nephrology, ed 5. Philadelphia, Lippincott Williams & Wilkins, 2004, pp 643–645.

31 Koyama A, Inage H, Kobayashi M, Nakamura H, Narita M, Tojo S: Effect of chemical cationization of antigen on glomerular localization of immune complexes in active models of serum sickness nephritis in rabbits. Immunology 1986;58:529–534.
32 Chen JS, Chen A, Chang LC, et al: Mouse model of membranous nephropathy induced by cationic bovine serum albumin: antigen dose-response relations and strain differences. Nephrol Dial Transplant 2004;19:2721–2728.
33 Davis PJ, Williams SC: Protein modification by thermal processing. Allergy 1998;53:102–105.
34 Sanchez C, Fremont S: Consequences of heat treatment and processing of food on the structure and allergenicity of component proteins. Rev Fr Allergol Immunol Clin 2003;43:13–20.
35 Sathe SK, Teuber SS, Roux KH: Effects of food processing on the stability of food allergens. Biotechnol Adv 2005;23:423–429.
36 Sreedharan R, Mehta DI: Gastrointestinal tract. Pediatrics 2004;113:1044–1050.
37 Alting AC, Meijer RJ, van Beresteijn EC: Incomplete elimination of the ABBOS epitope of bovine serum albumin under simulated gastrointestinal conditions of infants. Diabetes Care 1997;20:875–880.
38 Schmidt DG, Meijer RJ, Slangen CJ, van Beresteijn EC: Raising the pH of the pepsin-catalysed hydrolysis of bovine whey proteins increases the antigenicity of the hydrolysates. Clin Exp Allergy 1995;25:1007–1017.
39 Van Elburg RM, Fetter WP, Bunkers CM, Heymans HS: Intestinal permeability in relation to birth weight and gestational and postnatal age. Arch Dis Child Fetal Neonatal Ed 2003;88:F52–F55.

Pierre Ronco
INSERM Unit 702, Hôpital Tenon
4 rue de la Chine
FR–75020 Paris (France)
E-Mail pierre.ronco@tnn.aphp.fr

Liu Z-H, He JC (eds): Podocytopathy. Contrib Nephrol. Basel, Karger, 2014, vol 183, pp 157–173
DOI: 10.1159/000360053

Podocytopathy in Diabetes: A Metabolic Disorder

Sandra Merscher[a–c] • Oliver Lenz[a] • Alessia Fornoni[a–c]

[a]Division of Nephrology and Hypertension, Department of Medicine and [b]Diabetes Research Institute, and [c]Peggy and Harold Katz Family Drug Discovery Center, University of Miami Miller School of Medicine, Miami, Fla., USA

Abstract

Background: Despite multifactorial interventions, the prevalence of diabetic kidney disease (DKD) is reaching epidemic proportions worldwide. Loss of podocytes, the cells forming the glomerular filtration barrier, occurs early in the development of DKD and is associated with the presence of albuminuria. ***Summary:*** Recent studies have implicated circulating hormones, adipokines and lipids in the development and progression of DKD. ***Key Message:*** Although the endocrine effects of these factors are widely recognized, less is known about their direct effects on podocytes and the resulting implications for DKD.

Diabetic kidney disease (DKD) is a worldwide epidemic and the most common cause of end-stage renal disease (ESRD) in the United States. The prevalence of DKD has increased in parallel with the increased prevalence of diabetes, and despite higher usage rates of glucose-lowering medications and antagonists of the renin-aldosterone system, the prevalence of DKD among those with diabetes has remained unchanged [1]. Chronic kidney disease has been linked to increased mortality [2, 3], and consequently early interventions to treat DKD are important. Characteristic histopathologic findings of DKD include glomerular and tubular basement membrane (GBM) thickening, mesangial expansion leading to Kimmelstiel-Wilson nodules, glomerular and arteriolar hyalinosis, and the loss of podocytes (podocytopenia).

Podocytes are specialized, differentiated cells of the kidney glomerulus, which consist of a cell body, major processes and foot processes. Podocyte foot processes are linked to the GBM via their actin cytoskeleton and foot processes from neighboring cells form an interdigitating pattern leaving filtration slits between them. These slits are bridged by the slit diaphragm, and together they play an important role in the selective permeability of the glomerular filtration barrier [4–6]. Integrity of the glomerular filtration barrier is key to preventing the loss of protein into the urine (proteinuria), and research has shown that mutations in genes coding for slit-diaphragm proteins cause proteinuria-associated congenital nephropathies [7–11].

Microalbuminuria is the first clinical manifestation of DKD and inversely correlates with the number of podocytes in experimental animal models and in humans [12–18]. Podocytopenia has been recognized as a key event in the development of DKD in patients with diabetes [13, 15–18].

Research in the past decade has shown that podocytes express a variety of functional receptors making them direct targets of metabolic and endocrine factors (fig. 1). Additionally, clinical trials have shown that multifactorial intervention mainly aimed at targeting hyperglycemia, hypertension and hyperlipidemia can delay but not stop the progression of DKD [19–21]. This suggests that other mediators, such as metabolic and endocrine factors, may play an important role in disease development and/or progression.

Due to space limitations, this review will focus on podocytes as targets of metabolic and endocrine factors in DKD without commenting on the role of endothelial and mesangial cells in the process of disease progression.

Renin-Angiotensin-Aldosterone System

The renin-angiotensin system (RAS) is an important regulator of cardiovascular and renal function. Local RAS systems have been described in different tissues, including the kidney and podocytes, which express all key components of the system, including angiotensinogen, renin, ACE1, AT1 and AT2 [22–24], as well as the less traditional components ACE2 [23] and prorenin receptors [25]. In conditions resembling the diabetic milieu, the expression of several RAS components ultimately leads to podocyte apoptosis [26–28]. Consistent, animal models with activated RAS develop proteinuria with pathological features of DKD, while RAS blockage ameliorates proteinuria and delays the progression of DKD [29, 30]. Clinical studies demonstrated that angiotensin-converting enzyme (ACE) inhibitors or ARBs reduce proteinuria in DKD, although this did not always translate into a decreased risk of ESRD. Thus, the anti-proteinuric effect cannot be explained by the blood pressure-lowering actions of RAS blockage alone [31–33], but may result from a direct effect of RAS components on renal cells.

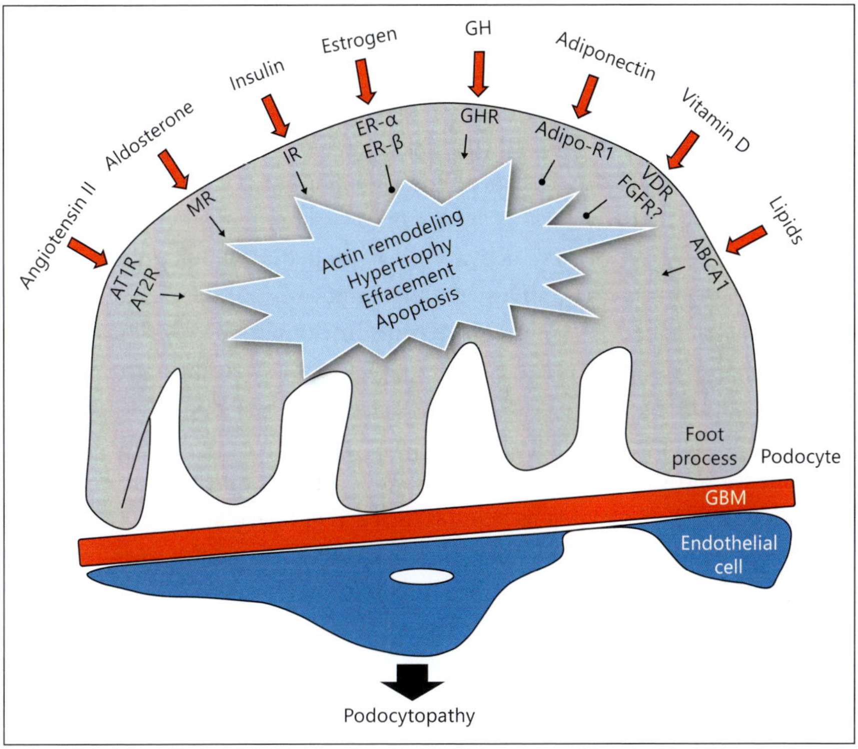

Fig. 1. Causes of podocytopathy in diabetes. Podocytes express a variety of functional receptors making them direct targets of metabolic and endocrine factors. AT1R = Angiotensin receptor type 1; AT2R = angiotensin receptor type 2; GH = growth hormone; MR = mineralocorticoid receptor; IR = insulin receptor; ER = estrogen receptor; GHR = growth hormone receptor; Adipo-R1 = adiponectin receptor 1; VDR = vitamin D receptor; FGFR = fibroblast growth factor receptor; ABCA1 = ATP-binding cassette transporter A1; GBM = glomerular basement membrane. Black lines with arrowheads indicate activating; black lines with circles indicate inhibitory effects.

Plasma prorenin levels are elevated 50–200% in diabetic patients with albuminuria [34] and have been shown to precede the development of microalbuminuria in patients with type 1 diabetes [35]. Non-proteolytic activation of prorenin has been suggested as a mechanism to activate the renal RAS, thus contributing to the development of proteinuria and glomerulosclerosis in hypertensive rats [36]. In contrast to the other components of the RAS system, overexpression of ACE2 under the nephrin promoter in transgenic mice delays and ameliorates the development of the glomerular phenotype in STZ-injected mice. Glomerular protection occurred in the absence of changes in blood pressure pointing to a blood pressure-independent local effect of ACE2 on the glomerulus [29].

Aldosterone

More recently, much attention has been given to the association of aldosterone, a steroid hormone with mineralocorticoid activity, and proteinuria. Mineral-corticoid receptors (MR) are expressed in glomerular podocytes [37], and aldosterone-mediated renal tissue injury is associated with increased reactive oxygen species levels [37, 38], activation of mitogen-activated protein kinases [39], and Rho kinases [40], and increased apopotosis [39, 41]. In uninephrectomized rats on a high-salt diet, aldosterone infustion induced proteinuria and hypertension [38, 42], and as selective blockage of aldosterone improved podocyte damage and proteinuria in animal models [37] and in patients with type 1 and type 2 diabetes [43]. Based on the fact that aldosterone has MR-independent effects and that MR may be activated by ligands other than aldosterone [reviewed in 44], it remains to be established if glomerular protection will be more likely achieved with MR antagonism, aldosterone-synthase inhibition, or both.

Insulin

More than 30 years ago it was suggested that the kidney is insulin-sensitive based on the observation that the infusion of insulin in patients caused dose-dependent albuminuria despite normoglycemia. Since then, research has shown that within the kidney glomerulus, podocytes express several glucose transporters (1, 2, 3, 4 and 8) that are activated in diabetes [45–48] as well as insulin receptor substrate-1 and insulin receptor (IR) [47]. Insulin resistance correlates with the development of microalbuminuria in patients with type 1 or type 2 diabetes [49–52], their siblings [53, 54], and in subjects without diabetes [55], and insulin-sensitizing agents such as thiazolidinediones exert a renoprotective function in humans [56]. In addition, patients with genetic mutations in the IR [57] and podocyte-specific IR knockout mice develop a DKD-like phenotype in the absence of hyperglycemia [58, 59], suggesting an important role for insulin signaling in podocyte function that does not depend on hyperglycemia. Insulin signaling is impaired in podocytes isolated from experimental models of diabetic nephropathy prior to the onset of microalbuminuria [60]. A similar impairment is observed in glomeruli isolated from diabetic rats [61]. In addition, several factors such as nephrin [62], the lipid phosphatase SHIP2 [63], and septin 7 [64] may have an important role in linking insulin signaling to the actin cytoskeleton in podocytes. The direct effect of insulin on podocyte depolarization [65, 66] and VEGF production

[67] further support a role of insulin as an important endocrine modulator of podocyte function. The role of glucose transporters is more complex. Overexpression of GLUT1 in mesangial cells leads to a phenotype that resembles DKD [68]. This is not the case for podocytes, where podocyte-specific overexpression of GLUT1 prevents mesangial expansion [69], suggesting the presence of cell type-specific functions of GLUTs. Finally, the ability of IGFs such IGF-II [70] to affect podocyte survival warrants further investigation.

Estrogen

Estrogens, especially 17β-estradiol (E_2), play an important role in attenuating glomerulosclerosis, tubulointerstitial fibrosis, and proteinuria in clinical and experimental models of DKD [71–77]. In support of this observation, the female to male incidence ratio of ESRD due to DKD increases after the age of menopause [78]. Podocytes express estrogen receptors α and β (ER-α and ER-β), making the glomerulus a target tissue for estrogen action [71]. Activation of ER-α protects podocytes from puromycin-induced apoptosis [79]. In db/db mice, E_2 treatment protected podocytes from injury and improved diabetes-induced kidney disease in association with decreased TGF-β and increased ER-β expression [80]. E_2 deficiency increases the severity of the renal phenotype in STZ-treated diabetic rats [74], while E_2 replacement attenuates glomerular and tubulointerstitial disease in aging estrogen-deficient Dahl salt-sensitive rats [76]. ER-α knockout mice are characterized by reduced podocyte numbers compared to controls underlining an important role of ER-α in podocyte viability and function [79]. In addition, podocytes isolated from E_2-treated mice show higher expression of estrogen receptors, suggesting a direct protective effect of E_2. E_2 supplementation leads to decreased TGF-β expression, decreased accumulation of collagen type IV, fibronectin, and laminin, and increased expression of matrix metalloproteinases I and II [71–73, 75–77]. Furthermore, a direct effect of E_2 on podocyte actin cytoskeleton composition and small GTPase activity has been reported [80].

Growth Hormone/Growth Hormone Receptor

Growth hormone (GH) is a 191-amino-acid, single-chain polypeptide. GH is synthesized and secreted in the anterior pituitary gland and it stimulates growth, cell reproduction and regeneration. Alterations in the GH/growth hormone receptor (GHR) axis have been described in diabetes and DKD [81–84]. A direct effect of GH on podocytes in DKD has been suggested given that podocytes express GHRs [85], and because exposure of podocytes to GH resulted in altered intracellular redistribution of the JAK-2 adapter protein Src homology 2-Bb (SH2-Bb), activation of focal adhesion kinase (FAK), increased reactive oxygen species, and actin cytoskeleton remodeling. Overexpression of bovine GH in transgenic mice has been linked to DKD [82, 86], whereas mice with a disrupted GH receptor/binding protein gene [87] or those expressing a bovine GH antagonist were protected from STZ-induced glomerular hypertrophy [86]. Clinical studies in children and adolescents with type 1 diabetes confirmed a potential role of renal GH in the pathogenesis of early DKD [81], underlining the importance of the GH/GHR axis in podocytes.

Adiponectin

Adipocytokines are peptides that are secreted by adipose tissue with adiponectin being the most abundantly secreted. Adiponectin is an adipose-specific collagen-like factor [88, 89] with insulin-sensitizing, anti-inflammatory, and vascular protective functions. Adiponectin exists in three subforms of low, medium, and high molecular weight. The high molecular weight subform is thought to be the biologically active form [90, 91]. Interestingly, adiponectin and the adiponectin receptor, Adipo-R1, are expressed in podocytes and contribute to the regulation of NADPH oxidase (Nox4) through the 5′-adenosine monophosphate-activated protein pathway [92, 93], making the podocyte a target for adiponectin action. Decreased circulating adiponectin levels are observed in obese patients and in patients with atherosclerosis and insulin resistance, and decreased levels correlate with cardiovascular risk in patients without renal disease [94, 95]. In contrast, increased adiponectin levels are associated with progression to ESRD in patients with type 1 and type 2 diabetes [95–99] and in mouse models of chronic renal failure [100]. Adiponectin knockout mice are characterized by albuminuria, a phenotype that was reversed by treatment of these mice with adiponectin, establishing for the first time a causative link between adiponectin and albuminuria [93]. It appears that increased adiponectin levels in ESRD patients or in models with chronic renal failure may be the consequence of decreased clearance rather than representing a cause for ESRD. A single nucleotide polymorphism in the promoter region of adiponectin was described to be associated with DKD patients with type 1 diabetes [101], indicating an important role of adiponectin in the development of DKD.

Vitamin D and Fibroblast Growth Factors

Vitamin D is a fat-soluble vitamin that helps maintain serum calcium and phosphate concentrations and ensures normal bone growth, remodeling, and mineralization. Additional functions include regulation of cell growth, anti-inflammatory properties, and neuromuscular- and immune-related functions. Most of the circulating vitamin D is synthesized in skin exposed to UV light, hydroxylated first in the liver to 25(OH)D_3 and then in the kidney to the active form 1,25$(OH)_2D_3$ [102]. Podocytes express vitamin D receptor (VDR) [103], which is highly inducible by 1,25$(OH)_2D_3$, and 1,25$(OH)_2D_3$ prevents high glucose-induced apoptosis in podocytes [104]. Interestingly, it was shown that 1,25$(OH)_2D_3$ also inhibits renin gene transcription and angiotensinogen expression in mesangial cells and podocytes by blocking the NF-κB pathway [105], suggesting a protective role of vitamin D in DKD via inhibition of the RAS system. In support of this protective role, VDR transgenic mice are partially protected from STZ-induced DKD and albuminuria, and podocyte-specific reconstitution of VDR signaling in VDR null mice with a hVDR transgene rescues VDR-null mice from severe diabetes-related renal damage [104]. In humans, low levels of vitamin D are associated with CKD progression and mortality [106]. Reconstitution of vitamin D levels reduces albuminuria and decreases cardiovascular mortality in patients with CKD [107–110]. Thus, targeting VDR signaling in podocytes may represent a new approach to prevent or treat DKD. Fibroblast growth factors (FGF) such as FGF-23 modulate renal hydroxylation of vitamin D and may mediate complications of CKD via this pathway [111]. However, vitamin D-independent effects of FGF will have to be explored as well, given that FGF-2 infusion in rats causes glomerulosclerosis [112], and FGF receptors such as R4 are expressed in kidney biopsies of patients with glomerular disorders [113].

Lipids

Dyslipidemia is a characteristic feature of DKD and serum lipids are thought to predict cardiac death in patients with type 1 and type 2 diabetes [114–116]. Current recommendations to target hyperlipidemia include statin therapy for low density lipoprotein cholesterol reduction [117]. Although statin therapy has proven to reduce the risk of cardiovascular disease in diabetic individuals [118, 119], the beneficial effect of statin therapy on albuminuria and kidney disease remains controversial, particularly in patients with ESRD [120–123].

It has been suggested that the accumulation of cholesterol and/or other lipid metabolites within cells rather than in the serum may contribute to DKD and other kidney diseases [124–130]. As a physiological concentration of intracellular cholesterol is essential for proper insulin signaling [131, 132], altered cholesterol distribution under pathological conditions may cause insulin resistance as described in adipocytes [133]. In diabetes, an altered cellular lipid balance was described in adipocytes where intracellular cholesterol accumulation was associated with cellular insulin resistance [134], a condition that also causes podocyte malfunction as described above.

In addition, accumulation of lipids other than cholesterol such as ceramides, GM3 gangliosides, and diacylglycerol in skeletal muscle cells, hepatocytes, and adipocytes has been described and is associated with insulin resistance [135–142]. Increased accumulation of cholesterol was described in glomeruli of mouse models of type 1 and type 2 diabetes [143–145]. Altered distribution of cholesterol is also characteristic of several disease conditions of genetic [146] and non-genetic origin [133, 147].

Increased cellular cholesterol content can result from increased influx, increased synthesis, or decreased efflux. Circulating low density lipoproteins are the major cholesterol carrier involved in cholesterol uptake in a receptor-dependent manner [148], while ABC transporters play a key role in cholesterol efflux to HDL acceptors [149]. Niemann-Pick C1 and C2 proteins play an important role in intracellular cholesterol transport from the lysosomes to the endoplasmic reticulum [150, 151]. Key enzymes regulating cholesterol synthesis are 3-hydroxy-3-methyl-glutaryl-CoA reductase, which is the direct target of statin action and cholesterol acyltransferase, and which in turn are regulated by sterol-regulated membrane-bound transcription factors [152, 153].

Using a cell-based assay in which differentiated human podocytes were exposed to sera from patients with DKD, we showed that cellular cholesterol accumulation in podocytes occurred in association with decreased expression of ATP-binding cassette transporter 1 *(ABCA1)* [154]. Likewise, we and others showed that in mouse models for type 1 and type 2 diabetes, accumulation of cholesterol in the kidney occurs due to decreased *ABCA1* expression rather than increased cholesterol synthesis [144, 154]. In agreement with this observation, the capacity of serum from patients with type 2 diabetes to induce cellular cholesterol efflux is impaired [155], and we showed that targeting reverse cholesterol transport with cyclodextrin rather than targeting cholesterol synthesis with statins may prove beneficial in preserving podocyte function and viability [154]. Therefore, targeting reverse cholesterol transport may open a new avenue for the treatment of DKD.

Conclusion

Podocyte injury in diabetes can result from endocrine, paracrine, and autocrine metabolic derangements that are not necessarily due to hyperglycemia and that may result from a complex cross-talk between different organs.

Disclosure Statement

The authors have no conflicts of interest to disclose.

Acknowledgements

S.M. and A.F. are supported by the US National Institutes of Health (DK090316), by the Diabetes Research Institute Foundation, by the Nephcure Foundation, by the Peggy and Harold Katz Family Foundation, and by the Diabetic Complications Consortium (U24DK076169-08). S.M. is supported by Stanley J. Glaser Foundation Research Award. Part of this project was supported by Grant No. 1UL1TR000460, University of Miami Clinical and Translational Science Institute, from the National Center for Advancing Translational Sciences and the National Institute on Minority Health and Health Disparities.

References

1 De Boer IH, Rue TC, Hall YN, Heagerty PJ, Weiss NS, Himmelfarb J: Temporal trends in the prevalence of diabetic kidney disease in the United States. JAMA 2011;305:2532–2539.

2 Federation ID: IDF Diabetes Atlas Update 2012. IDF Diabetes Atlas, ed 5. Brussels, International Diabetes Federation, 2011.

3 System USRD: USRDS 2012 Annual Data Report: Atlas of Chronic Kidney Disease and End-Stage Renal Disease in the United States. Bethesda, National Institutes of Health, National Institute of Diabetes and Digestive and Kidney Diseases, 2012.

4 Pavenstadt H, Kriz W, Kretzler M: Cell biology of the glomerular podocyte. Physiol Rev 2003;83:253–307.

5 Somlo S, Mundel P: Getting a foothold in nephrotic syndrome. Nat Genet 2000;24:333–335.

6 Faul C, Asanuma K, Yanagida-Asanuma E, Kim K, Mundel P: Actin up: regulation of podocyte structure and function by components of the actin cytoskeleton. Trends Cell Biol 2007;17:428–437.

7 Kestila M, Lenkkeri U, Mannikko M, Lamerdin J, McCready P, Putaala H, Ruotsalainen V, Morita T, Nissinen M, Herva R, et al: Positionally cloned gene for a novel glomerular protein – nephrin – is mutated in congenital nephrotic syndrome. Mol Cell 1998;1:575–582.

8 Boute N, Gribouval O, Roselli S, Benessy F, Lee H, Fuchshuber A, Dahan K, Gubler MC, Niaudet P, Antignac C: NPHS2, encoding the glomerular protein podocin, is mutated in autosomal recessive steroid-resistant nephrotic syndrome. Nat Genet 2000;24:349–354.

48 Lewko B, Bryl E, Witkowski JM, Latawiec E, Golos M, Endlich N, Hahnel B, Koksch C, Angielski S, Kriz W, Stepinski J: Characterization of glucose uptake by cultured rat podocytes. Kidney Blood Press Res 2005;28:1–7.

49 Ekstrand AV, Groop PH, Gronhagen-Riska C: Insulin resistance precedes microalbuminuria in patients with insulin-dependent diabetes mellitus. Nephrol Dial Transplant 1998; 13:3079–3083.

50 Yip J, Mattock MB, Morocutti A, Sethi M, Trevisan R, Viberti G: Insulin resistance in insulin-dependent diabetic patients with microalbuminuria. Lancet 1993;342:883–887.

51 Groop L, Ekstrand A, Forsblom C, Widen E, Groop PH, Teppo AM, Eriksson J: Insulin resistance, hypertension and microalbuminuria in patients with type 2 (non-insulin-dependent) diabetes mellitus. Diabetologia 1993;36:642–647.

52 Parvanova AI, Trevisan R, Iliev IP, Dimitrov BD, Vedovato M, Tiengo A, Remuzzi G, Ruggenenti P: Insulin resistance and microalbuminuria: a cross-sectional, case-control study of 158 patients with type 2 diabetes and different degrees of urinary albumin excretion. Diabetes 2006;55:1456–1462.

53 Forsblom CM, Eriksson JG, Ekstrand AV, Teppo AM, Taskinen MR, Groop LC: Insulin resistance and abnormal albumin excretion in non-diabetic first-degree relatives of patients with NIDDM. Diabetologia 1995;38: 363–369.

54 Yip J, Mattock M, Sethi M, Morocutti A, Viberti G: Insulin resistance in family members of insulin-dependent diabetic patients with microalbuminuria. Lancet 1993;341:369–370.

55 Mykkanen L, Zaccaro DJ, Wagenknecht LE, Robbins DC, Gabriel M, Haffner SM: Microalbuminuria is associated with insulin resistance in nondiabetic subjects: the insulin resistance atherosclerosis study. Diabetes 1998; 47:793–800.

56 Miyazaki Y, Cersosimo E, Triplitt C, DeFronzo RA: Rosiglitazone decreases albuminuria in type 2 diabetic patients. Kidney Int 2007; 72:1367–1373.

57 Musso C, Javor E, Cochran E, Balow JE, Gorden P: Spectrum of renal diseases associated with extreme forms of insulin resistance. Clin J Am Soc Nephrol 2006;1:616–622.

58 Welsh GI, Hale LJ, Eremina V, Jeansson M, Maezawa Y, Lennon R, Pons DA, Owen RJ, Satchell SC, Miles MJ, et al: Insulin signaling to the glomerular podocyte is critical for normal kidney function. Cell Metab 2010;12: 329–340.

59 Fornoni A: Proteinuria, the podocyte, and insulin resistance. N Engl J Med 2011;363: 2068–2069.

60 Tejada T, Catanuto P, Ijaz A, Santos JV, Xia X, Sanchez P, Sanabria N, Lenz O, Elliot SJ, Fornoni A: Failure to phosphorylate AKT in podocytes from mice with early diabetic nephropathy promotes cell death. Kidney Int 2008;73:1385–1393.

61 Mima A, Ohshiro Y, Kitada M, Matsumoto M, Geraldes P, Li C, Li Q, White GS, Cahill C, Rask-Madsen C, King GL: Glomerular-specific protein kinase C-β-induced insulin receptor substrate-1 dysfunction and insulin resistance in rat models of diabetes and obesity. Kidney Int 2011;79:883–896.

62 Coward RJ, Welsh GI, Koziell A, Hussain S, Lennon R, Ni L, Tavare JM, Mathieson PW, Saleem MA: Nephrin is critical for the action of insulin on human glomerular podocytes. Diabetes 2007;56:1127–1135.

63 Hyvonen ME, Saurus P, Wasik A, Heikkila E, Havana M, Trokovic R, Saleem M, Holthofer H, Lehtonen S: Lipid phosphatase SHIP2 downregulates insulin signalling in podocytes. Mol Cell Endocrinol 2010;328:70–79.

64 Wasik AA, Polianskyte-Prause Z, Dong MQ, Shaw AS, Yates JR 3rd, Farquhar MG, Lehtonen S: Septin-7 forms a complex with CD2AP and nephrin and regulates glucose transporter trafficking. Mol Biol Cell 2012;23: 3370–3379.

65 Kim EY, Anderson M, Dryer SE: Insulin increases surface expression of TRPC6 channels in podocytes: role of NADPH oxidases and reactive oxygen species. Am J Physiol Renal Physiol 2012;302:F298–F307.

66 Kim EY, Dryer SE: Effects of insulin and high glucose on mobilization of slo1 BKCa channels in podocytes. J Cell Physiol 2011;226: 2307–2315.

67 Hale LJ, Hurcombe J, Lay A, Santamaria B, Valverde AM, Saleem MA, Mathieson PW, Welsh GI, Coward RJ: Insulin directly stimulates VEGF-A production in the glomerular podocyte. Am J Physiol Renal Physiol 2013; 305:F182–F188.

68 Wang Y, Heilig K, Saunders T, Minto A, Deb DK, Chang A, Brosius F, Monteiro C, Heilig CW: Transgenic overexpression of GLUT1 in mouse glomeruli produces renal disease resembling diabetic glomerulosclerosis. Am J Physiol Renal Physiol 2010;299:F99–F111.
69 Zhang H, Schin M, Saha J, Burke K, Holzman LB, Filipiak W, Saunders T, Xiang M, Heilig CW, Brosius FC 3rd: Podocyte-specific overexpression of GLUT1 surprisingly reduces mesangial matrix expansion in diabetic nephropathy in mice. Am J Physiol Renal Physiol 2010;299:F91–F98.
70 Hale LJ, Welsh GI, Perks CM, Hurcombe JA, Moore S, Hers I, Saleem MA, Mathieson PW, Murphy AJ, Jeansson M, et al: Insulin-like growth factor-II is produced by, signals to and is an important survival factor for the mature podocyte in man and mouse. J Pathol 2013;230:95–106.
71 Catanuto P, Doublier S, Lupia E, Fornoni A, Berho M, Karl M, Striker GE, Xia X, Elliot S: 17β-Estradiol and tamoxifen upregulate estrogen receptor β expression and control podocyte signaling pathways in a model of type 2 diabetes. Kidney Int 2009;75:1194–1201.
72 Dixon A, Wells CC, Singh S, Babayan R, Maric C: Renoprotective effects of a selective estrogen receptor modulator, raloxifene, in an animal model of diabetic nephropathy. Am J Nephrol 2007;27:120–128.
73 Keck M, Romero-Aleshire MJ, Cai Q, Hoyer PB, Brooks HL: Hormonal status affects the progression of STZ-induced diabetes and diabetic renal damage in the VCD mouse model of menopause. Am J Physiol Renal Physiol 2007;293:F193–F199.
74 Mankhey RW, Bhatti F, Maric C: 17β-Estradiol replacement improves renal function and pathology associated with diabetic nephropathy. Am J Physiol Renal Physiol 2005;288:F399–F405.
75 Mankhey RW, Wells CC, Bhatti F, Maric C: 17β-Estradiol supplementation reduces tubulointerstitial fibrosis by increasing MMP activity in the diabetic kidney. Am J Physiol Regul Integr Comp Physiol 2007;292:R769–R777.
76 Maric C, Sandberg K, Hinojosa-Laborde C: Glomerulosclerosis and tubulointerstitial fibrosis are attenuated with 17β-estradiol in the aging Dahl salt-sensitive rat. J Am Soc Nephrol 2004;15:1546–1556.
77 Wells CC, Riazi S, Mankhey RW, Bhatti F, Ecelbarger C, Maric C: Diabetic nephropathy is associated with decreased circulating estradiol levels and imbalance in the expression of renal estrogen receptors. Gend Med 2005;2:227–237.
78 USRDS: 2011 Annual Data Report: Atlas of Chronic Kidney Disease and End-Stage Renal Disease in the United States. Bethesda, National Institutes of Health, National Institute of Diabetes and Digestive and Kidney Diseases, 2011.
79 Kummer S, Jeruschke S, Wegerich LV, Peters A, Lehmann P, Seibt A, Mueller F, Koleganova N, Halbenz E, Schmitt CP, et al: Estrogen receptor alpha expression in podocytes mediates protection against apoptosis in vitro and in vivo. PLoS One 2011;6:e27457.
80 Catanuto P, Fornoni A, Pereira-Simon S, Wu F, Burnstein KL, Xia X, Conti F, Lenzi A, Elliot S: In vivo 17β-estradiol treatment contributes to podocyte actin stabilization in female db/db mice. Endocrinology 2012;153:5888–5895.
81 Cummings EA, Sochett EB, Dekker MG, Lawson ML, Daneman D: Contribution of growth hormone and IGF-I to early diabetic nephropathy in type 1 diabetes. Diabetes 1998;47:1341–1346.
82 Doi T, Striker LJ, Quaife C, Conti FG, Palmiter R, Behringer R, Brinster R, Striker GE: Progressive glomerulosclerosis develops in transgenic mice chronically expressing growth hormone and growth hormone releasing factor but not in those expressing insulin-like growth factor-1. Am J Pathol 1988;131:398–403.
83 Karl M, Potier M, Schulman IH, Rivera A, Werner H, Fornoni A, Elliot SJ: Autocrine activation of the local insulin-like growth factor I system is upregulated by estrogen receptor (ER)-independent estrogen actions and accounts for decreased ER expression in type 2 diabetic mesangial cells. Endocrinology 2005;146:889–900.
84 Tack I, Elliot SJ, Potier M, Rivera A, Striker GE, Striker LJ: Autocrine activation of the IGF-I signaling pathway in mesangial cells isolated from diabetic NOD mice. Diabetes 2002;51:182–188.

85 Reddy GR, Pushpanathan MJ, Ransom RF, Holzman LB, Brosius FC 3rd, Diakonova M, Mathieson P, Saleem MA, List EO, Kopchick JJ, et al: Identification of the glomerular podocyte as a target for growth hormone action. Endocrinology 2007;148:2045–2055.

86 Chen NY, Chen WY, Bellush L, Yang CW, Striker LJ, Striker GE, Kopchick JJ: Effects of streptozotocin treatment in growth hormone and growth hormone antagonist transgenic mice. Endocrinology 1995;136:660–667.

87 Bellush LL, Doublier S, Holland AN, Striker LJ, Striker GE, Kopchick JJ: Protection against diabetes-induced nephropathy in growth hormone receptor/binding protein gene-disrupted mice. Endocrinology 2000; 141:163–168.

88 Nakano Y, Tobe T, Choi-Miura NH, Mazda T, Tomita M: Isolation and characterization of GBP28, a novel gelatin-binding protein purified from human plasma. J Biochem 1996;120:803–812.

89 Maeda K, Okubo K, Shimomura I, Funahashi T, Matsuzawa Y, Matsubara K: cDNA cloning and expression of a novel adipose-specific collagen-like factor, apM1 (AdiPose Most abundant Gene transcript 1). Biochem Biophys Res Commun 1996;221:286–289.

90 Pajvani UB, Du X, Combs TP, Berg AH, Rajala MW, Schulthess T, Engel J, Brownlee M, Scherer PE: Structure-function studies of the adipocyte-secreted hormone Acrp30/adiponectin. Implications fpr metabolic regulation and bioactivity. J Biol Chem 2003;278:9073–9085.

91 Pajvani UB, Hawkins M, Combs TP, Rajala MW, Doebber T, Berger JP, Wagner JA, Wu M, Knopps A, Xiang AH, et al: Complex distribution, not absolute amount of adiponectin, correlates with thiazolidinedione-mediated improvement in insulin sensitivity. J Biol Chem 2004;279:12152–12162.

92 Cammisotto PG, Bendayan M: Adiponectin stimulates phosphorylation of AMP-activated protein kinase α in renal glomeruli. J Mol Histol 2008;39:579–584.

93 Sharma K, Ramachandrarao S, Qiu G, Usui HK, Zhu Y, Dunn SR, Ouedraogo R, Hough K, McCue P, Chan L, et al: Adiponectin regulates albuminuria and podocyte function in mice. J Clin Invest 2008;118:1645–1656.

94 Guebre-Egziabher F, Bernhard J, Funahashi T, Hadj-Aissa A, Fouque D: Adiponectin in chronic kidney disease is related more to metabolic disturbances than to decline in renal function. Nephrol Dial Transplant 2005;20:129–134.

95 Frystyk J, Tarnow L, Hansen TK, Parving HH, Flyvbjerg A: Increased serum adiponectin levels in type 1 diabetic patients with microvascular complications. Diabetologia 2005;48:1911–1918.

96 Jorsal A, Tarnow L, Frystyk J, Lajer M, Flyvbjerg A, Parving HH, Vionnet N, Rossing P: Serum adiponectin predicts all-cause mortality and end-stage renal disease in patients with type 1 diabetes and diabetic nephropathy. Kidney Int 2008;74:649–654.

97 Kato K, Osawa H, Ochi M, Kusunoki Y, Ebisui O, Ohno K, Ohashi J, Shimizu I, Fujii Y, Tanimoto M, Makino H: Serum total and high molecular weight adiponectin levels are correlated with the severity of diabetic retinopathy and nephropathy. Clin Endocrinol (Oxf) 2008;68:442–449.

98 Lin J, Hu FB, Curhan G: Serum adiponectin and renal dysfunction in men with type 2 diabetes. Diabetes Care 2007;30:239–244.

99 Saraheimo M, Forsblom C, Fagerudd J, Teppo AM, Pettersson-Fernholm K, Frystyk J, Flyvbjerg A, Groop PH: Serum adiponectin is increased in type 1 diabetic patients with nephropathy. Diabetes Care 2005;28:1410–1414.

100 Komura N, Kihara S, Sonoda M, Maeda N, Tochino Y, Funahashi T, Shimomura I: Increment and impairment of adiponectin in renal failure. Cardiovasc Res 2010;86:471–477.

101 Zhang D, Ma J, Brismar K, Efendic S, Gu HF: A single nucleotide polymorphism alters the sequence of SP1 binding site in the adiponectin promoter region and is associated with diabetic nephropathy among type 1 diabetic patients in the Genetics of Kidneys in Diabetes Study. J Diabetes Complications 2009;23:265–272.

102 Holick MF: Vitamin D deficiency. N Engl J Med 2007;357:266–281.

103 Wang Y, Borchert ML, Deluca HF: Identification of the vitamin D receptor in various cells of the mouse kidney. Kidney Int 2012; 81:993–1001.

104 Wang Y, Deb DK, Zhang Z, Sun T, Liu W, Yoon D, Kong J, Chen Y, Chang A, Li YC: Vitamin D receptor signaling in podocytes protects against diabetic nephropathy. J Am Soc Nephrol 2012;23:1977–1986.
105 Deb DK, Chen Y, Zhang Z, Zhang Y, Szeto FL, Wong KE, Kong J, Li YC: 1,25-Dihydroxyvitamin D_3 suppresses high glucose-induced angiotensinogen expression in kidney cells by blocking the NF-κB pathway. Am J Physiol Renal Physiol 2009;296:F1212–F1218.
106 Ravani P, Malberti F, Tripepi G, Pecchini P, Cutrupi S, Pizzini P, Mallamaci F, Zoccali C: Vitamin D levels and patient outcome in chronic kidney disease. Kidney Int 2009;75:88–95.
107 Kovesdy CP, Ahmadzadeh S, Anderson JE, Kalantar-Zadeh K: Association of activated vitamin D treatment and mortality in chronic kidney disease. Arch Intern Med 2008;168:397–403.
108 Kuhlmann A, Haas CS, Gross ML, Reulbach U, Holzinger M, Schwarz U, Ritz E, Amann K: 1,25-Dihydroxyvitamin D_3 decreases podocyte loss and podocyte hypertrophy in the subtotally nephrectomized rat. Am J Physiol Renal Physiol 2004;286:F526–F533.
109 Shoben AB, Rudser KD, de Boer IH, Young B, Kestenbaum B: Association of oral calcitriol with improved survival in nondialyzed CKD. J Am Soc Nephrol 2008;19:1613–1619.
110 Tentori F, Hunt WC, Stidley CA, Rohrscheib MR, Bedrick EJ, Meyer KB, Johnson HK, Zager PG: Mortality risk among hemodialysis patients receiving different vitamin D analogs. Kidney Int 2006;70:1858–1865.
111 Thadhani R, Wolf M: Vitamin D in patients with kidney disease: cautiously optimistic. Adv Chronic Kidney Dis 2007;14:22–26.
112 Kriz W, Hahnel B, Rosener S, Elger M: Long-term treatment of rats with FGF-2 results in focal segmental glomerulosclerosis. Kidney Int 1995;48:1435–1450.
113 Zhu CP, Yi ZW: Expression of fibroblast growth factor receptor-4 in fetal kidneys and pathological kidneys of children (in Chinese) Zhongguo Dang Dai Er Ke Za Zhi 2007;9:133–138.
114 Tschope W, Koch M, Thomas B, Ritz E: Serum lipids predict cardiac death in diabetic patients on maintenance hemodialysis. Results of a prospective study. The German Study Group Diabetes and Uremia. Nephron 1993;64:354–358.
115 Makinen VP, Soininen P, Kangas AJ, Forsblom C, Tolonen N, Thorn LM, Viikari J, Raitakari OT, Savolainen M, Groop PH, Ala-Korpela M: Triglyceride-cholesterol imbalance across lipoprotein subclasses predicts diabetic kidney disease and mortality in type 1 diabetes: the FinnDiane Study. J Intern Med 2013;273:383–395.
116 Thomas MC, Rosengard-Barlund M, Mills V, Ronnback M, Thomas S, Forsblom C, Cooper ME, Taskinen MR, Viberti G, Groop PH: Serum lipids and the progression of nephropathy in type 1 diabetes. Diabetes Care 2006;29:317–322.
117 Standards of Medical Care in Diabetes – 2011. Diabetes Care 2011;34(suppl 1):S11–S61.
118 Colhoun HM, Betteridge DJ, Durrington PN, Hitman GA, Neil HA, Livingstone SJ, Thomason MJ, Mackness MI, Charlton-Menys V, Fuller JH: Primary prevention of cardiovascular disease with atorvastatin in type 2 diabetes in the Collaborative Atorvastatin Diabetes Study (CARDS): multicentre randomised placebo-controlled trial. Lancet 2004;364:685–696.
119 MRC/BHF Heart Protection Study of cholesterol lowering with simvastatin in 20,536 high-risk individuals: a randomised placebo-controlled trial. Lancet 2002;360:7–22.
120 Colhoun HM, Betteridge DJ, Durrington PN, Hitman GA, Neil HA, Livingstone SJ, Charlton-Menys V, DeMicco DA, Fuller JH: Effects of atorvastatin on kidney outcomes and cardiovascular disease in patients with diabetes: an analysis from the Collaborative Atorvastatin Diabetes Study (CARDS). Am J Kidney Dis 2009;54:810–819.
121 Abe M, Maruyama N, Okada K, Matsumoto S, Matsumoto K, Soma M: Effects of lipid-lowering therapy with rosuvastatin on kidney function and oxidative stress in patients with diabetic nephropathy. J Atheroscler Thromb 2011;18:1018–1028.

122 Atthobari J, Brantsma AH, Gansevoort RT, Visser ST, Asselbergs FW, van Gilst WH, de Jong PE, de Jong-van den Berg LT: The effect of statins on urinary albumin excretion and glomerular filtration rate: results from both a randomized clinical trial and an observational cohort study. Nephrol Dial Transplant 2006;21:3106–3114.
123 Tonolo G, Velussi M, Brocco E, Abaterusso C, Carraro A, Morgia G, Satta A, Faedda R, Abhyankar A, Luthman H, Nosadini R: Simvastatin maintains steady patterns of GFR and improves AER and expression of slit diaphragm proteins in type 2 diabetes. Kidney Int 2006;70:177–186.
124 Geoffroy K, Wiernsperger N, Lagarde M, El Bawab S: Bimodal effect of advanced glycation end products on mesangial cell proliferation is mediated by neutral ceramidase regulation and endogenous sphingolipids. J Biol Chem 2004;279:34343–34352.
125 Hao CM, Breyer MD: Roles of lipid mediators in kidney injury. Semin Nephrol 2007; 27:338–351.
126 Hao CM, Breyer MD: Physiologic and pathophysiologic roles of lipid mediators in the kidney. Kidney Int 2007;71:1105–1115.
127 Lennon R, Pons D, Sabin MA, Wei C, Shield JP, Coward RJ, Tavare JM, Mathieson PW, Saleem MA, Welsh GI: Saturated fatty acids induce insulin resistance in human podocytes: implications for diabetic nephropathy. Nephrol Dial Transplant 2009;24:3288–3296.
128 Mather AR, Siskind LJ: Glycosphingolipids and kidney disease. Adv Exp Med Biol 2011; 721:121–138.
129 Zador IZ, Deshmukh GD, Kunkel R, Johnson K, Radin NS, Shayman JA: A role for glycosphingolipid accumulation in the renal hypertrophy of streptozotocin-induced diabetes mellitus. J Clin Invest 1993;91:797–803.
130 Geoffroy K, Troncy L, Wiernsperger N, Lagarde M, El Bawab S: Glomerular proliferation during early stages of diabetic nephropathy is associated with local increase of sphingosine-1-phosphate levels. FEBS Lett 2005;579:1249–1254.
131 Saltiel AR, Pessin JE: Insulin signaling in microdomains of the plasma membrane. Traffic 2003;4:711–716.
132 Watson RT, Shigematsu S, Chiang SH, Mora S, Kanzaki M, Macara IG, Saltiel AR, Pessin JE: Lipid raft microdomain compartmentalization of TC10 is required for insulin signaling and GLUT4 translocation. J Cell Biol 2001;154:829–840.
133 Le Lay S, Krief S, Farnier C, Lefrere I, Le Liepvre X, Bazin R, Ferre P, Dugail I: Cholesterol, a cell size-dependent signal that regulates glucose metabolism and gene expression in adipocytes. J Biol Chem 2001; 276:16904–16910.
134 Yu BL, Zhao SP, Hu JR: Cholesterol imbalance in adipocytes: a possible mechanism of adipocytes dysfunction in obesity. Obes Rev 2010;11:560–567.
135 Holland WL, Bikman BT, Wang LP, Yuguang G, Sargent KM, Bulchand S, Knotts TA, Shui G, Clegg DJ, Wenk MR, et al: Lipid-induced insulin resistance mediated by the proinflammatory receptor TLR4 requires saturated fatty acid-induced ceramide biosynthesis in mice. J Clin Invest 2011;121:1858–1870.
136 Holland WL, Brozinick JT, Wang LP, Hawkins ED, Sargent KM, Liu Y, Narra K, Hoehn KL, Knotts TA, Siesky A, et al: Inhibition of ceramide synthesis ameliorates glucocorticoid-, saturated-fat-, and obesity-induced insulin resistance. Cell Metab 2007; 5:167–179.
137 Holland WL, Knotts TA, Chavez JA, Wang LP, Hoehn KL, Summers SA: Lipid mediators of insulin resistance. Nutr Rev 2007; 65:S39–S46.
138 Holland WL, Summers SA: Sphingolipids, insulin resistance, and metabolic disease: new insights from in vivo manipulation of sphingolipid metabolism. Endocr Rev 2008; 29:381–402.
139 Hunnicutt JW, Hardy RW, Williford J, McDonald JM: Saturated fatty acid-induced insulin resistance in rat adipocytes. Diabetes 1994;43:540–545.
140 Itani SI, Ruderman NB, Schmieder F, Boden G: Lipid-induced insulin resistance in human muscle is associated with changes in diacylglycerol, protein kinase C, and IκB-α. Diabetes 2002;51:2005–2011.

141 Montell E, Turini M, Marotta M, Roberts M, Noe V, Ciudad CJ, Mace K, Gomez-Foix AM: DAG accumulation from saturated fatty acids desensitizes insulin stimulation of glucose uptake in muscle cells. Am J Physiol Endocrinol Metab 2001;280:E229–E237.

142 Schmitz-Peiffer C: Targeting ceramide synthesis to reverse insulin resistance. Diabetes 2010;59:2351–2353.

143 Proctor G, Jiang T, Iwahashi M, Wang Z, Li J, Levi M: Regulation of renal fatty acid and cholesterol metabolism, inflammation, and fibrosis in Akita and OVE26 mice with type 1 diabetes. Diabetes 2006;55:2502–2509.

144 Tang C, Kanter JE, Bornfeldt KE, Leboeuf RC, Oram JF: Diabetes reduces the cholesterol exporter ABCA1 in mouse macrophages and kidneys. J Lipid Res 2010;51: 1719–1728.

145 Wang Z, Jiang T, Li J, Proctor G, McManaman JL, Lucia S, Chua S, Levi M: Regulation of renal lipid metabolism, lipid accumulation, and glomerulosclerosis in FVBdb/db mice with type 2 diabetes. Diabetes 2005;54: 2328–2335.

146 Peake KB, Vance JE: Defective cholesterol trafficking in Niemann-Pick C-deficient cells. FEBS Lett 2010;584:2731–2739.

147 Khera AV, Cuchel M, de la Llera-Moya M, Rodrigues A, Burke MF, Jafri K, French BC, Phillips JA, Mucksavage ML, Wilensky RL, et al: Cholesterol efflux capacity, high-density lipoprotein function, and atherosclerosis. N Engl J Med 2011;364:127–135.

148 Goldstein JL, Brown MS: The LDL receptor. Arterioscler Thromb Vasc Biol 2009;29: 431–438.

149 Fielding CJ, Fielding PE: Cellular cholesterol efflux. Biochim Biophys Acta 2001;1533: 175–189.

150 Abi-Mosleh L, Infante RE, Radhakrishnan A, Goldstein JL, Brown MS: Cyclodextrin overcomes deficient lysosome-to-endoplasmic reticulum transport of cholesterol in Niemann-Pick type C cells. Proc Natl Acad Sci USA 2009;106:19316–19321.

151 Maxfield FR, Wustner D: Intracellular cholesterol transport. J Clin Invest 2002;110: 891–898.

152 Brown MS, Goldstein JL: A proteolytic pathway that controls the cholesterol content of membranes, cells, and blood. Proc Natl Acad Sci USA 1999;96:11041–11048.

153 Gong Y, Lee JN, Lee PC, Goldstein JL, Brown MS, Ye J: Sterol-regulated ubiquitination and degradation of Insig-1 creates a convergent mechanism for feedback control of cholesterol synthesis and uptake. Cell Metab 2006;3:15–24.

154 Merscher-Gomez S, Guzman J, Pedigo CE, Lehto M, Aguillon-Prada R, Mendez A, Lassenius MI, Forsblom C, Yoo T, Villarreal R, et al: Cyclodextrin protects podocytes in diabetic kidney disease. Diabetes 2013;62: 3817–3827.

155 Zhou H, Tan KC, Shiu SW, Wong Y: Cellular cholesterol efflux to serum is impaired in diabetic nephropathy. Diabetes Metab Res Rev 2008;24:617–623.

Alessia Fornoni, MD, PhD
Division of Nephrology and Hypertension
University of Miami
1580 NW 10th Avenue, Miami, FL 33136 (USA)
E-Mail AFornoni@med.miami.edu

Liu Z-H, He JC (eds): Podocytopathy. Contrib Nephrol. Basel, Karger, 2014, vol 183, pp 174–180
DOI: 10.1159/000360109

Obesity-Related Glomerulopathy and Podocyte Injury

Marcello Camici
University Hospitals, Pisa, Italy

Abstract

Obesity-related glomerulopathy is a dysfunction of podocytes present in obese proteinuric patients that may evolve in progressive loss of renal function. It is morphologically defined as focal segmental glomerulosclerosis and glomerulomegaly characterized by podocyte hypertrophy with a reduced density. The article reviews and discusses the pathogenetic mechanisms of the podocyte injury during obesity and proposes possible therapeutic strategies.

Obesity-related glomerulopathy (ORG) is a secondary form of focal segmental glomerulosclerosis where glomerular visceral epithelial cells contribute to the formation of glomerular crescents and where glomerulomegaly is present. Decreased podocyte density and number are present in ORG; these changes in podocytes correlate with the degree of proteinuria and loss of renal function in obese patients [1]. The incidence of ORG has increased in recent years as observed in the study of Kambham et al. [2] on renal biopsies. Hsu et al. [3] show in a cohort study that ORG is a risk factor for end-stage renal disease. Despite these observations and numerous publications, reviews and editorials about ORG, the pathogenetic mechanism of obesity-induced glomerulomegaly and focal segmental glomerulosclerosis is incompletely understood. We discuss the pathogenetic mechanisms and propose possible targeting pharmacological interventions of podocyte injury in ORG.

Pathogenetic Mechanisms

Renin-Angiotensin System

Obesity has an elevated glomerular filtration rate (obesity, hyperfiltration) associated with a less constant increase in renal plasma flow, resulting in a variable increase in filtration fraction. Glomerular hyperfiltration may induce glomerulomegaly and glomerular hypertension. The pathogenesis of these glomerular hemodynamic abnormalities that contribute to the podocyte injury play an important role in the obesity renin-angiotensin activation induced by multiple factors, including secretion of renin and renin precursors by adipocytes [4]. Circulating levels of angiotensin II (AngII) increase with increasing body mass index and AngII receptors are present in podocytes. They respond to AngII with an increase in free cytosolic calcium resulting in a depolarization of podocyte cell [5]. There is also experimental evidence that AngII directly modulates the podocyte function via transient receptor potential canonical channel-6 (TRCP-6) located in the slit diaphragm; podocyte TRCP-6 overexpression in acquired human disease is sufficient to induce podocyte actin reorganization and proteinuria [6].

Aldosterone

Adipose tissue is capable of stimulating AngII-independent aldosterone (Ald) production. Complement-C1q TNF-related protein-1 (CTRP-1), a novel adipokine, stimulates Ald production in male Sprague-Dawley rats in the zona glomerulosa of the adrenal cortex [7]. Mineral corticoid receptors have been detected in the podocytes: their activation by Ald induces an effect on kinase SgK1, activates nicotinamide adenine dinucleotide phosphate oxidase (NADPH) and generates reactive oxygen species [8]. Ald receptor blockers show antifibrotic effects in patients with chronic kidney disease [9], Ald antagonism attenuates obesity-induced hypertension and glomerular hyperfiltration [10], and finally in transgenic Ren2 rats the podocyte foot process effacement is normalized by treatment with spironolactone [11].

Plasminogen Activator Inhibitor-1 (PAI-1)

In obesity the PAI-1 physiological inhibitor of plasminogen activators is increased in adipose tissue and glomerular cells where the action of free fatty acids on adipose tissue macrophages dynamically regulates the expression of PAI-1 [12]. PAI-1 modulates adipocyte differentiation [13] and is expressed in glomerulonephritis [14]. As the primary physiological inhibitor of plasminogen activators, PAI-1 plays a role in kidney fibrosis [15]. Prevention of obesity and insulin resistance is possible in mice lacking PAI-1 [16]. In a podocyte injury-associated glomerulosclerosis model, renoprotection conferred by peroxisome proliferator-activated receptor-γ agonist is achieved, in part, through decreased PAI-1 [17]. These observations suggest that PAI-1 may modulate podocyte injury during obesity. Much still remains to be understood about the role of PAI-1 in kidney disease and in particular its role in renal fibrogenesis promotion.

Adiponectin

Adiponectin (Adipo) is an adipokine secreted mainly by adipocytes – it is a protein with insulin-sensitizing, anti-inflammatory, anti-atherogenic, immunodepressant and vasculoprotective properties. These biological functions are exploited by coupling to the receptors R1/R2. Adipo-R1 receptor is responsible for the AMPK (5-adenosine monophosphate protein kinase) signaling pathway activation, whereas Adipo-R2 receptor is responsible for PPAR-α activation [18]. Glomerular cells express Adipo-R1 receptor and Adipo null mutant mice show podocyte foot process fusion, oxidant stress and albuminuria. Treatment with Adipo normalizes podocyte effacement and albuminuria [19]. Adipo stimulation of the AMPK pathway suppresses an isoform of NADPH (Nox-4) and therefore oxidative stress. Blood Adipo levels are depressed in obesity [20] and such a situation triggers in podocyte cells oxidative stress via the Nox-4 isoform.

Lipid Metabolism

Hyperlipidemia may contribute to renal damage and saturated fatty acids induce insulin resistance in human podocytes. Obese hyperlipidemic mice (young C57BL/6 mice) induced by a high-fat diet develop ORG [21]. Lipids such as oxidized LDL lead to redistribution and loss of nephrin as well as podocyte apoptosis by decreasing phosphorylation of Akt, a prominent pathway for cell survival [22], and statin protects against podocyte apoptosis in vitro. Sterol regulatory element binding protein-1 (SREBP-1) appears to play a role in lipid renal accumulation: diet-induced obesity in C57B/LY mice causes increased renal accumulation and glomerulosclerosis via a SREBP-1-dependent pathway [23]. In a rodent model of high-fat diet-induced proteinuria and glomerular disease, the farnesoid X receptor modulated lipid metabolism and diet-induced renal inflammation, fibrosis and proteinuria [24].

Macrophages

Obesity-related macrophage infiltration of adipose tissue plays a role in the low-grade chronic inflammatory state present in obesity. Depending on the local microenvironment and stage of tissue injury, macrophages display heterogeneity in functions [25]. Obesity induces a macrophage phenotypic switch in adipose tissue characterized by production of proinflammatory cytokines,

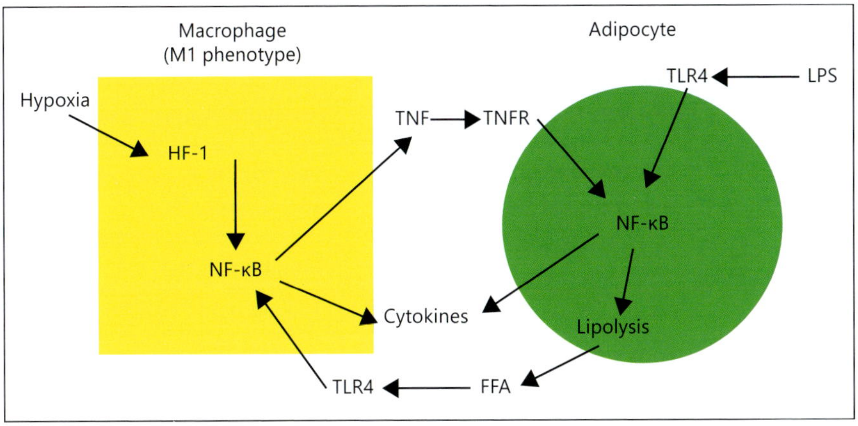

Fig. 1. Local cross-talk between adipose tissue macrophages and adipocytes in white adipose tissue during obesity (left side: macrophage (M1 phenotype polarization); right side: adipocyte). In adipose tissue the hypoxia stimulates via hypoxia-inducible factor 1 (HF-1) the activation of nuclear factor-κB (NF-κB) signaling and adipose tissue macrophages secrete high amounts of tumor necrosis factor (TNF)-stimulating receptor (TNF-R) on adipocytes that induces NF-κB signaling pathway activation. The TNF-induced NF-κB activation in adipocytes results in increased lipolysis with release from adipocytes of free fatty acids (FFA). These FFA act as paracrine TLR4 ligands that induce NF-κB signaling pathway activation inside macrophages. The macrophage FFA-activated NF-κB signaling pathway stimulates and maintains TNF secretion and proinflammatory cytokine release. Adipocytes express TLR4 and stimulation by lipopolysaccharide (LPS) induces proinflammatory cytokine secretion by adipocytes.

chemokines and cellular adhesion molecules (fig. 1). In cultured mouse podocytes the activated macrophages downregulate podocyte nephrin and podocin expression whereas in conditionally immortalized podocytes the macrophages suppress the activity of nephrin gene promoter. In cultured rat of glomerular epithelial cells, TNF-α induces actin cytoskeleton reorganization. Toll-like receptor 4 (TLR4) links podocytes with the innate immune system.

Oxidative Stress

The nephropathy of Zucker diabetic fat rat provides evidence for the role of oxidative and nitrosoactive stress [26] and another experimental model of obesity such as the spontaneously hypertensive/NIH-corpulent rat SHR/NDmcr-cp(cp/cp) also provides such evidence [27]. Heme oxygenase plays a key role in renal function regulation [28] by attenuating the production of reactive oxygen species because degradation products of heme oxygenase possess potent antioxidant and antiapoptic activities, and obesity is associated with a decrease in heme oxygenase type 1.

Table 1. Focal segmental glomerulosclerosis in obesity: interventions based on podocyte injury

General mechanism	Specific intervention	
	currently available	possible in future
RAAS	ACE inhibitors AngII receptor block Inhibition of Ald receptors	Selective TRCP-6 inhibitors AngI receptor block Inhibition of non-genomic effect of Ald Inhibition CTRP-1
PAI-1	AngII inhibitors	Selective PAI-1 inhibitors
Lipids	Body weight reduction Statin	Ligands of farnesoid X receptor
Adiponectin	Body weight reduction RAAS blockade Enhancement of Adipo intracellular signal pathway	Adipo administration Agonist Adipo R1/2 receptors
Macrophages	Steroids	Selective inhibitors of TLR4 Selective block of inflammatory kinases
Oxidative stress	Antioxidants	HO-1 overexpression Inhibition of specific NADPH oxidase (Nox-4 isoform) Peroxynitrate scavenger
Glomerular hyperfiltration	Salt restriction RAAS blockade	Endothelin receptor blockade

ACE = Angiotensin-converting enzyme; HO-1 = heme oxygenase-1; CTRP1 = complement-C1q TNF-related protein-1.

Pharmacological Interventions Perspective

Current available therapy may be used for counteracting pathogenetic mechanisms of ORG. Based on the mechanism of podocyte injury, the improving knowledge about the pathogenetic mechanisms of this injury in ORG suggest development of new antiproteinuric therapies. The current available therapy may be associated with possible future specific pharmacological interventions (table 1).

We suggest specific interventions of inhibition such as selective TRCP-6 inhibitors or AT1 receptor, CTRP-1 and non-genomic effects of Ald inhibition. Selective PAI-1 inhibitors may be useful for the control of renal profibrotic effects of PAI-1 and ligands of farnesoid X receptor for the control of renal toxicity of lipids. Adipo administration for increasing blood levels of

Adipo in ORG or agonists of Adipo-R1 receptor may counteract podocyte injury produced by hypoadiponectinemia. Selective inhibitors of TLR4 receptor and inflammatory kinases of the intracellular signaling pathway may be developed for counteracting the proinflammatory adipose macrophage. Peroxynitrate scavenger, inhibition of specific NADPH, such as the Nox-4 isoform, and heme oxigenase-1 overexpression may be useful for decreasing oxidative stress. Finally, specific interventions in renal hemodynamics beyond dietary sodium restriction and/or renin-angiotensin-aldosterone system (RAAS) blockade deserve further exploration; the role of insulin resistance and endothelins as factors contributing to glomerular hemodynamic changes remains to be clarified.

Conclusions

Glomerular visceral epithelial cells, also called podocytes, cover the external surface of the glomerular basement membrane in humans and animals. Podocytes contribute to various functions in the glomerulus, in particular they serve as the final filtration barrier to prevent the leak of plasma proteins. The pathogenetic mechanisms discussed are experimental, i.e. derived from in vitro research and from a proteinuric experimental animal model study. Pharmacological research of pathogenetic mechanisms, in particular beyond Adipo in the adipokine field such as leptin, resistin, and visfatin, must be developed. It is widely accepted that proteinuria reduction is an imperative therapeutic goal in chronic proteinuric kidney disease. Proteinuria is a strong risk factor for the progression of chronic kidney disease: during obesity it is a hallmark of ORG and the podocyte injury is a common threat for proteinuric disease.

References

1 Chen HM, Liu ZH, Zeng CI, Li SH, Wang QW: Podocyte lesions in patients with obesity-related glomerulopathy. Am J Kidney Dis 2006;48:772–779.

2 Kambham N, Markowitz GS, Valeri AM, Lin J, D'Agati VD: Obesity-related glomerulopathy: an emerging epidemic. Kidney Int 2001; 59:1498–1509.

3 Hsu CY, McCulloch CE, Iribarren C: Body mass index and risk for end-stage renal disease. Ann Intern Med 2006;144:21–28.

4 Gorzelniak K, Engeli S, Janke J, et al: Hormonal regulation of the human adipose-tissue renin-angiotensin system: relationship to obesity and hypertension. J Hypertens 2002; 20:965–973.

5 Pavenstad H, Kritz W, Kretzler M: Cell biology of the glomerular podocyte. Physiol Rev 2003;83:253–307.

6 Moller CC, Wei C, Altintas MM: Induction of TRCP-6 channels in acquired forms of proteinuria kidney disease. J Am Soc Nephrol 2007;18:29–36.

7 Jeon JH, Kim KY, Kim JK, Baek A, Cho H, Kim JW, Han SH, Lim JS, Yang Y: A novel adipokine CTRP-1 stimulates aldosterone production. FASEB J 2008;22:1502–1511.
8 Nagase M, Fujita T: Aldosterone and glomerular podocyte injury. Clin Exp Nephrol 2008; 12:233–242.
9 Guney I: Antifibrotic effects of aldosterone receptor blocker (spironolactone) in patients with chronic kidney disease. Renal Failure 2009;31:779–784.
10 De Paula RB, Da Silva AA, Hall JE: Aldosterone antagonism attenuates obesity-induced hypertension and glomerular hyperfiltration. Hypertension 2004;43:41–47.
11 Whaley-Connell A, Habibi J, Wei Y: Mineralocorticoid receptor antagonism attenuates glomerular filtration barrier remodeling in the transgenic Ren2 rat. Am J Physiol Renal Physiol 2009;296:F1013–F1022.
12 Eddy AA, Fogo AB: Plasminogen activator inhibitor-1 in chronic kidney disease: evidence and mechanisms of action. J Am Soc Nephrol 2006;17:2999–3012.
13 Liang X, Kanjanabuch T, Mao SL: Plasminogen activator inhibitor-1 modulates adipocyte differentiation. Am J Physiol Endocrinol Metab 2006;290:E103–E113.
14 Lee HS, Park SY, Moon KC, Hong SY: mRNA expression of urokinase and plasminogen activator inhibitor-1 in human crescent glomerulonephritis. Histopathology 2001;39: 203–209.
15 Ma LJ, Fogo AB: PAI-1 and kidney fibrosis. Front Biosci 2009;14:2028–2041.
16 Ma LJ, Mao SL, Taylor KL: Prevention of obesity and insulin resistance in mice lacking plasminogen activator inhibitor-1. Diabetes 2004;53:336–346.
17 Yang HC, Ma LJ, Fogo AB: Peroxisome proliferator-activated receptor-γ agonist is protective in podocyte injury-associated sclerosis. Kidney Int 2006;69:1756–1764.
18 Kadowaki T, Yamauchi T, Kubota N: The physiological and pathophysiological role of adiponectin receptors in the peripheral tissues and CNS. FEBS Lett 2008;582:579–584.
19 Sharma K, Ramachadraro S, Qiu G: Adiponectin regulates albuminuria and podocyte function in mice. J Clin Invest 2008;118: 1645–1656.
20 Yatagai T: Hypoadiponectinemia is associated with visceral fat accumulation and insulin resistance in Japanese men with type 1 diabetes mellitus. Metabolism 2003;52:1274–1278.
21 Deji N, Kume S, Araki S: Structural and functional changes in the kidneys of high-fat diet-induced obese mice. Am J Physiol Renal Physiol 2009;296:F118–F126.
22 Bussolati B, Deregibus MC, Fonsato V: Statins prevent oxidized LDL-induced injury of glomerular podocytes by activating the phosphatidylinositol 3-kinase/Akt-signaling pathway. J Am Soc Nephrol 2005;16:1936–1947.
23 Jiang T, Wang Z, Proctor G: Diet-induced obesity in C57BL/6J mice causes increased renal accumulation and glomerulosclerosis via a sterol regulatory element-binding protein-1-dependent pathway. J Biol Chem 2005;280: 32317–32395.
24 Wang XX, Jiang T, Shen Y: The farnesoid X receptor modulates renal lipid metabolism and diet-induced renal inflammation, fibrosis and proteinuria. Am J Physiol Renal Physiol 2009;297:F1587–F1596.
25 Ricardo SD, van Goor H, Eddy AA: Macrophage diversity in renal injury and repair. J Clin Invest 2008;118:3522–3530.
26 Chander PN, Gealekman O, Brodsky SV: Nephropathy in Zucker diabetic fat rat is associated with oxidative and nitrososactive stress: prevention by chronic therapy with a peroxynitrate scavenger ebselen. J Am Soc Nephrol 2004;15:2391–2403.
27 Nangaku M, Izuhara Y, Usuda N: Type 2 diabetic nephropathy rat model, the improvement of obesity by a low caloric diet reduces oxidative/carbonyl stress and prevents diabetic nephropathy. Nephrol Dial Transpl 2005;20:2661–2669.
28 Abraham NG, Cao J, Sacerdoti D, Li X, Drummond G: Heme oxygenase: the key to renal function regulation. Am J Physiol Renal Physiol 2009;297:F1137–F1152.

Marcello Camici
University Hospitals
Via Roma n. 65
IT–56100 Pisa (Italy)
E-Mail mcamici@tiscalinet.it

Liu Z-H, He JC (eds): Podocytopathy. Contrib Nephrol. Basel, Karger, 2014, vol 183, pp 181–190
DOI: 10.1159/000360113

Podocyte Injury in HIV-Associated Nephropathy

Sandeep K. Mallipattu[a] • John C. He[b, c]

[a]Division of Nephrology, Department of Medicine, Stony Brook University School of Medicine, [b]Division of Nephrology, Department of Medicine, Icahn School of Medicine at Mount Sinai, and [c]Renal Section, James J. Peters VA Medical Center, New York, N.Y., USA

Abstract

Prior to the widespread use of antiretroviral therapy, HIV-associated nephropathy (HIVAN) was the leading cause of end-stage renal disease among individuals of African descent. HIVAN, a rapidly progressing glomerulopathy, is characterized by collapsing focal segmental glomerulosclerosis, podocyte hypertrophy, and proliferation of epithelial cells overlying the Bowman's space. Although the origin of these proliferating cells is debatable, a large body of evidence suggests that podocyte dysfunction occurs in HIVAN. In the last decade, several groups have demonstrated that the podocyte serves as a reservoir for the virus and podocyte dysfunction is a direct result of specific viral protein expression. The viral protein, *nef*, directly activates several signaling pathways, resulting in podocyte dedifferentiation and proliferation. Today, with the widespread use of antiretroviral therapy, classical HIVAN is rare. Nonetheless, as patients with chronically infected HIV live longer, many still succumb to chronic kidney disease due to coexisting illnesses such as diabetes and hypertension. However, it remains unclear whether this viral reservoir serves a medium for development of non-HIVAN-related kidney diseases such as diabetic nephropathy. We also provide some insights into potential areas of research in characterizing podocyte biology in this changing spectrum of HIV-related kidney disease.

It has been more than 30 years since the first cases of acquired immunodeficiency syndrome (AIDS) were reported in the USA. Within a few years of those initial reported cases of AIDS, an aggressive glomerulopathy called HIV-associated nephropathy (HIVAN) was first diagnosed in the USA. HIVAN was the leading cause of end-stage renal disease (ESRD) among African-Americans in the early 1990s [1]. Histopathological findings of patients with HIVAN revealed a collapsing variant of focal segmental glomerulosclerosis (FSGS) with tubulointerstitial injury. Although the pathogenesis of HIVAN includes all

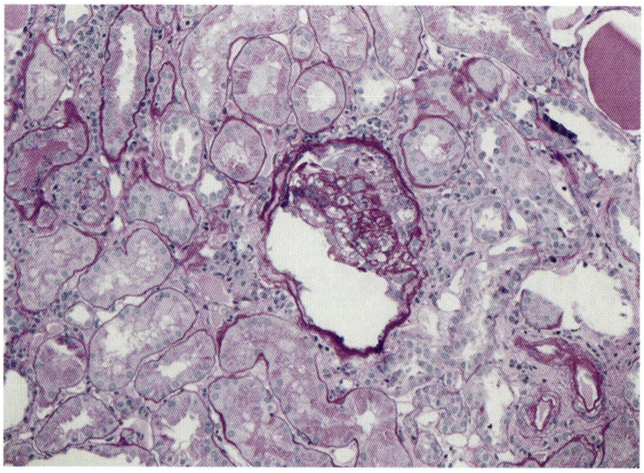

Fig. 1. Classical pathological features of HIVAN. Representative high-power image of a glomerulus demonstrating the collapse of capillary loops, dilatation of the glomerular space, and overlying podocyte hyperplasia. Periodic acid-Schiff. ×40.

compartments of the kidney, the glomerulus is the initial insult [2]. Specifically, the podocyte injury plays a vital role in the onset of albuminuria and progression of the disease. In this review, we provide an overview of the specific role of the podocyte in the pathogenesis and progression of HIVAN.

Histopathological Findings of Podocyte Injury

Renal biopsies of patients with HIVAN are typically consistent with collapsing glomerulopathy with tubulocystic dilatation. Specifically in the glomerulus, there is marked podocyte hypertrophy with foot process effacement and microvillus transformation. Ultrastructurally, podocytes exhibit enlarged vesicular nuclei, mitotic figures, and intracytoplasmic protein resorption droplets [3]. In addition, epithelial cell proliferation overlying the glomerulus can be prominent, resulting in cellular pseudocrescent formation (fig. 1). Although the origin of these proliferating epithelial cells remains debatable, it is clear there is dysregulation of podocyte function and morphology in HIVAN.

Pathogenesis of Podocyte Injury in HIVAN

Current Murine Model for HIVAN

After more than 20 years since its initial description, the HIV-1 transgenic mouse model, Tg26, has been widely used by researchers to characterize the pathogenesis of HIVAN. The Tg26 mouse consists of a 3-kb gag-pol deleted HIV-1 genome, resulting in a loss of viral replication, under the control of the viral long terminal repeat [4]. The heterozygous mouse results in representative features of glomerular injury typically observed in human HIVAN beginning at 4–6 weeks of age. However, the homozygous mice die prior to breeding [4]. Interestingly, the histopathological characteristics of HIVAN were observed only on the FVB background. In 2005, to delineate the role of podocyte dysfunction in HIVAN, Zhong et al. [2] developed a gag-pol-env deleted HIV-1 transgenic mouse specifically expressing HIV-1 proviral DNA in podocytes using the nephrin promoter. Remarkably, the expression of HIV-1 transgene in the podocyte alone resulted in collapsing FSGS, interstitial fibrosis, and diffuse microcystic tubular lesions [2], suggesting the potential role of the dysregulated podocyte in inducing chronic interstitial changes in HIVAN.

The Podocyte Serves as a Viral Reservoir in the Pathogenesis of HIVAN

A large body of evidence suggests that podocyte dedifferentiation in HIVAN is a direct result of HIV-1 infection of the cells, and independent of the systemic viral load. For instance, initial reciprocal transplantation studies by Bruggeman et al. [5] in 1997 revealed that HIV-1 gene expression in the kidney is required for the development of HIVAN. Additional studies confirmed the presence of HIV-1 RNA and DNA in many compartments, including the podocyte, in kidney biopsies from patients with HIV and undetectable viral loads, thereby validating the notion that the kidney serves as a reservoir for the virus [6]. Furthermore, many studies have clearly shown that the HIV-infected renal epithelial cells in the glomerulus constitute a distinct viral compartment as compared to peripheral blood. For instance, studies by Bruggeman et al. [5] demonstrated that HIV-1 proviral DNA is present in epithelial cells in the glomerulus with an undetectable HIV RNA levels in the peripheral blood. Additional studies using phylogenetic comparative analysis confirmed the existence of the renal viral reservoir by demonstrating there is kidney-specific viral evolution [7]. In addition, case reports substantiated this notion of a distinct renal viral compartment by demonstrating that despite the resolution of HIVAN with antiretroviral therapy, there is a persistence reservoir of HIV transcription in the kidney [8]. In the setting of interrupted therapy, this can lead to rapid formation of infectious virons.

Until recently, the mechanism by which HIV-1 entered the renal epithelial cells was unclear. Although studies were performed in renal tubular epithelial cells, Chen et al. [9] demonstrated that there is direct transfer of viral RNA between CD4+ T cells and renal epithelial cells independent of CD4 and Env expression. These studies have yet to be demonstrated using cultured podocytes.

Podocyte Response to Injury in HIVAN

Podocytes in the mature glomerulus are regarded as terminally differentiated and quiescent cells with minimal capacity for replication. However, in the setting of injury induced by HIV-1, the podocyte responds with dedifferentiation and proliferation. Specifically, a loss of podocyte differentiation markers such as synaptopodin, podocin, and Wilms' tumor-1 (WT-1) has been observed [10]. In addition, there is an increase in proliferation markers such as Ki-67, cyclin A, and cyclin D1 with a loss of cyclin-dependent kinase inhibitors (p21 and p27), suggesting the podocyte has re-entered the cell cycle [10]. Furthermore, primary podocytes in culture from HIV-1 transgenic mice as well as immortalized cultured podocytes infected with HIV-1 demonstrate spontaneous proliferation and reduced contact inhibition. Interestingly, not all glomerular diseases resulting in podocyte injury result in a similar loss of these podocyte differentiation markers [10].

In the setting of HIV-1 infection in the podocytes, it appears that not all viral proteins are required to induce podocyte dysregulation. Although initial studies by Husain et al. suggested that only the expression of *nef* was required to induce the podocytopathy observed in HIVAN, more recent studies have demonstrated a similar pattern with *vpr*. In fact, in vitro and in vivo studies have demonstrated that simply the presence of *nef* or *vpr* gene was required to induce podocyte dedifferentiation and proliferation [11]. Furthermore, it was shown that the expression of *nef* and *vpr* accelerated podocyte injury, suggesting a potential synergism in podocyte dysregulation [11].

To further delineate the mechanism by which viral genes such as *nef* can induce podocyte dysregulation, one has to closely examine the functional domains along this 27-kDa *nef* protein. One of the functional domains of *nef* is a proline-rich motif, which has been implicated in playing a critical role in enabling the activation of SH3 domain on Src kinase by *nef*. With this direct binding, the activation of Src-dependent Stat3 and Ras-c-Raf-MAPK1,2 pathways commences, resulting in podocyte dedifferentiation and proliferation [12]. These findings are abrogated in the setting of mutated *nef* that lacks proline-rich motif. Similarly, the alteration of SH3 domain on Src kinase results in the attenuation of nef-induced podocytopathy. Also, the reduction in Stat3 activity resulted in amelioration of podocyte injury typically observed in HIV1 transgenic mice [13]. Interestingly, immunostaining of human biopsy specimens revealed that the activation of these pathways only occurred in patients with HIVAN, in comparison to other glomerular diseases [12].

Vascular endothelial growth factor A (VEGF) plays a critical role in podocyte pathobiology. The overexpression of VEGF alone in mice results in collapsing glomerulopathy, resembling the lesions observed in HIVAN [14]. In vitro

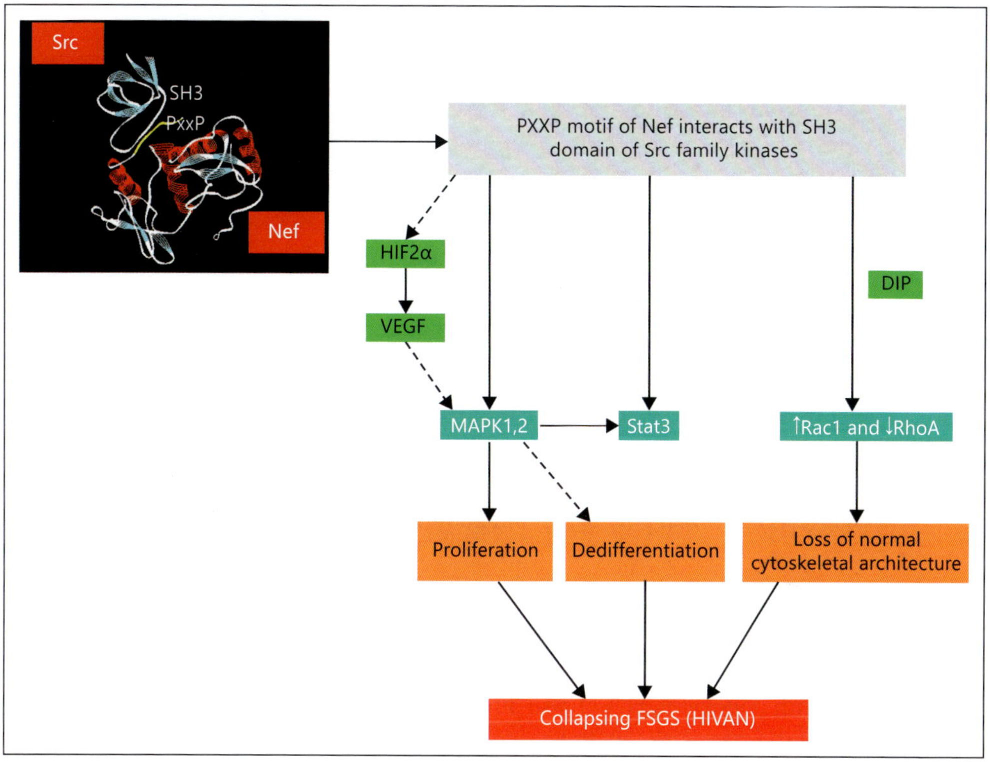

Fig. 2. Schematic diagram of interaction of *nef* in podocyte differentiation. The viral protein, *nef*, directly interacts with the SH3 domain of Src kinase, leading to activation of Stat3 and Ras-Raf MAPK1,2 pathways. This ultimately results in the loss of podocyte differentiation markers, thereby contributing to the collapsing glomerulopathy observed in HIVAN.

studies revealed that *nef* induces an increase in both VEGF expression as well as the activator of VEGF, hypoxia-inducible factor 2α (HIF2α). The upregulation of VEGF subsequently activates MAPK pathway, resulting in podocyte dedifferentiation and proliferation [15]. Furthermore, *nef*-induced HIF2α-VEGF is mediated by the Src-dependent STA3 pathway, further validating the critical role of *nef* in podocyte injury.

In addition to the activation of these pathways by *nef*, *nef* induces cytoskeletal changes observed in podocyte effacement by disrupting stress fibers and increasing lamellipodia, changes typically observed in HIVAN [16]. Specifically, *nef* induced Rac1 activation and RhoA inhibition via the recruitment of diaphanous interacting protein [16]. This alteration in RhoA/Rac1 results in cytoskeletal changes observed in HIVAN. A summary of *nef*-induced proliferation, dedifferentiation, and cytoskeletal changes observed in HIVAN is depicted in figure 2.

In general, a key component of tissue renewal involved the telomerase pathway. Telomerase is a critical enzyme complex involved in generating telomere repeats to prevent telomere dysfunction. A loss of telomere maintenance hinders cell survival and self-renewal. More recently, Shkreli et al. [17] demonstrated that merely the overexpression of telomerase reverse transcriptase (TERT) was required for induction of collapsing glomerulopathy typically observed in HIVAN. TERT is a co-factor involved in modulating the transcriptional responses regulated by Wnt signaling, a pathway previously demonstrated to contribute to podocyte dysfunction [17]. The overexpression of TERT activates the Wnt pathway, thereby resulting in podocyte dedifferentiation and proliferation. Furthermore, the authors demonstrated that this process can be reversed with restriction of transgenic TERT expression or inhibition of Wnt signaling pathway. This resulted in the podocyte exiting the cell cycle and re-expression of differentiation markers [17]. This study suggests again that podocyte proliferation could lead to collapsing FSGS.

What Is the Role of the Parietal Epithelial Cells in HIVAN?

As with podocytes, HIV-1 also infects parietal epithelial cells [18]. Recent literature raises the notion that the parietal epithelial cells may play a potential role in the replenishment of podocytes in the setting of podocyte injury. For instance, immunostaining studies reveal that cellular component of pseudocrescents observed in collapsing glomerulopathies, such as HIVAN, are primarily composed of cells positive for parietal epithelial cells markers [19]. Furthermore, in the podocyte-specific HIV-1 transgenic murine model, extracapillary proliferating cells stained positive for cytokeratin, but were void of podocyte markers, suggesting most of these proliferating cells were of parietal cell origin [2]. However, it remains unclear if these are truly proliferating parietal epithelial cells or merely podocytes that have lost their differentiation markers and have re-entered the cell cycle. Recent lineage tracing studies suggest that proliferative cells in crescents or pseudocrescents are from parietal epithelial cells [20]. However, lineage tracing studies have not been performed in HIV-1 transgenic mice to determine the origin of these proliferative cells observed in HIVAN.

Host Genetic Factors Play a Critical Role in HIVAN

Since HIVAN almost exclusively affects patients of African descent, host genetic factors likely play a critical role in the pathogenesis of HIVAN. Furthermore, the HIV-1 transgenic mouse model demonstrates a characteristic HIVAN phenotype on the FVB/n background, but is resistant to the phenotype on the C57BL/6 background [21]. Studies in recent years have focused on trying to identify novel factors and pathways that may explain this variance in phenotype [21–24]. For example, studies by Kaufman et al. [25] have demonstrated that an upregulation of mediators of cell adhesion, such as homophilic adhesion molecule sidekick-1 (sdk-1), may contribute to the clustering of podocytes observed in pseudocrescents seen in HIVAN lesions. In addition, the overexpression of sdk-1 resulted in severe FSGS and proteinuria by altering the slit-diaphragm linker protein, MAGI-1, thereby contributing to the disruption of actin cytoskeletal changes. Another example of the critical role of host genetic factors in HIVAN pathogenesis is the identification of HIVAN susceptibility loci. Using expression quantitative trait loci analysis, Papeta et al. [23] specifically demonstrated that the genetic lesions introduced by HIVAN susceptibility alleles alter podocyte regulatory network, thereby resulting in podocyte dysfunction and injury. More recently, the higher prevalence of HIVAN in patients of African descent may partly be explained by recently identified polymorphisms along chromosome 22q13, more specifically in apolipoprotein L-1 (APOL1) [26]. The hypothesized survival advantage of APOL1 variants over generations is attributed to the fact that variants of APOL1 appear to protect against African sleeping sickness as compared to wild-type APOL1 [26]. Furthermore, it is estimated that individuals with two APOL1 risk alleles exhibit a 50% lifetime risk of developing HIVAN [27]. Further conditional and haplotype analyses have validated that two APOL1 haplotypes accounted for almost all of the association with FSGS and HIVAN on chromosome 22q13 [24]. Interestingly, in normal human kidney specimens, APOL1 is exclusively present in the podocytes, with reduced expression in FSGS and HIVAN [28]. Further studies are required to determine whether the variants in APOL1 increase the susceptibility to HIVAN by potentiating *nef*-induced podocyte dedifferentiation and proliferation. Finally, therapy targeting the reversal of nef-induced podocyte dedifferentiation using retinoic acid has previously been described and will be the focus of an accompanying review in this edition by Sharma et al.

Podocyte Injury in the New Spectrum of HIV-Related Kidney Disease

Recent data from the USRDS reveals a plateau in the annual incidence of ESRD attributed to HIVAN, with approximately 800–900 new cases reported yearly [1]. Despite a plateau in incidence, the USRDS data demonstrates an increase in the prevalence of ESRD among HIV-infected individuals, which is likely a result of improved survival since the advent of antiretroviral since the mid-1990s.

Also, cross-sectional studies have confirmed that chronic kidney disease (CKD) in patients chronically infected with HIV is now more likely a result of non-HIVAN kidney disease in the USA [29]. As patients live longer with cART, the burden of CKD from coexisting illnesses such as diabetes mellitus or hypertension is likely to grow. The direct contribution of HIV infection to non-HIVAN kidney disease is not known, however available data suggest that HIV infection may have an independent effect on CKD progression. For instance, longitudinal studies have suggested a synergistic effect of HIV infection and diabetes in CKD progression [29]. Furthermore, the presence of a renal reservoir in patients chronically infected with HIV may provide a persistent low level of local inflammation, which may be exacerbated by coexisting illnesses such as diabetes mellitus and hypertension. For instance, using the resistant HIV-1 transgenic mouse on the C57BL/6 background, we demonstrated that diabetic HIV-1 transgenic mice exhibit a significant increase in histopathological features of diabetic nephropathy, including foot process effacement and a decrease in podocyte density [30]. Further studies are required to examine the mechanisms by which HIV-1 infection of the podocyte accelerates the progression of non-HIVAN-related CKD.

Conclusions

In conclusion, we provide a brief overview of the podocyte injury in the pathogenesis of HIVAN. Furthermore, we summarize the latest evidence in the role of genetic factors involved in the progression of podocyte injury in HIVAN. As patients live longer with chronically-infect HIV, the incidence of comorbid disease such as diabetes mellitus will likely increase. Therefore, future studies in HIV-related kidney diseases should focus on the mechanism by which the viral reservoir in the podocyte may exacerbate glomerulopathy secondary to coexisting chronic illnesses.

Acknowledgements

J.C.H. is supported by NIH 1R01DK078897 and Chinese 973 Fund 2012CB517601.

Disclosure Statement

The authors have no conflicts of interest to disclose.

References

1 United States Renal Data System: USRDS 2011 Annual Data Report: Atlas of Chronic Kidney Disease and End-Stage Renal Disease in the United States. http://www.usrds.org/adr.htm 2.

2 Zhong J, Zuo Y, Ma J, Fogo AB, Jolicoeur P, Ichikawa I, Matsusaka T: Expression of HIV-1 genes in podocytes alone can lead to the full spectrum of HIV-1-associated nephropathy. Kidney Int 2005;68:1048–1060.

3 D'Agati V, Appel GB: HIV infection and the kidney. J Am Soc Nephrol 1997;8:138–152.

4 Dickie P, Felser J, Eckhaus M, Bryant J, Silver J, Marinos N, Notkins AL: HIV-associated nephropathy in transgenic mice expressing HIV-1 genes. Virology 1991;185:109–119.

5 Bruggeman LA, Dikman S, Meng C, Quaggin SE, Coffman TM, Klotman PE: Nephropathy in human immunodeficiency virus-1 transgenic mice is due to renal transgene expression. J Clin Invest 1997;100:84–92.

6 Barisoni L, Bruggeman LA, Mundel P, D'Agati VD, Klotman PE: HIV-1 induces renal epithelial dedifferentiation in a transgenic model of HIV-associated nephropathy. Kidney Int 2000;58:173–181.

7 Marras D, Bruggeman LA, Gao F, Tanji N, Mansukhani MM, Cara A, Ross MD, Gusella GL, Benson G, D'Agati VD, Hahn BH, Klotman ME, Klotman PE: Replication and compartmentalization of HIV-1 in kidney epithelium of patients with HIV-associated nephropathy. Nat Med 2002;8:522–526.

8 Winston JA, Bruggeman LA, Ross MD, Jacobson J, Ross L, D'Agati VD, Klotman PE, Klotman ME: Nephropathy and establishment of a renal reservoir of HIV type 1 during primary infection. N Engl J Med 2001; 344:1979–1984.

9 Chen P, Chen BK, Mosoian A, Hays T, Ross MJ, Klotman PE, Klotman ME: Virological synapses allow HIV-1 uptake and gene expression in renal tubular epithelial cells. J Am Soc Nephrol 2011;22:496–507.

10 Barisoni L, Kriz W, Mundel P, D'Agati V: The dysregulated podocyte phenotype: a novel concept in the pathogenesis of collapsing idiopathic focal segmental glomerulosclerosis and HIV-associated nephropathy. J Am Soc Nephrol 1999;10:51–61.

11 Zuo Y, Matsusaka T, Zhong J, Ma J, Ma LJ, Hanna Z, Jolicoeur P, Fogo AB, Ichikawa I: HIV-1 genes *vpr* and *nef* synergistically damage podocytes, leading to glomerulosclerosis. J Am Soc Nephrol 2006;17:2832–2843.

12 He JC, Husain M, Sunamoto M, D'Agati VD, Klotman ME, Iyengar R, Klotman PE: Nef stimulates proliferation of glomerular podocytes through activation of Src-dependent Stat3 and MAPK1,2 pathways. J Clin Invest 2004;114:643–651.

13 Feng X, Lu TC, Chuang PY, Fang W, Ratnam K, Xiong H, Ouyang X, Shen Y, Levy DE, Hyink D, Klotman M, D'Agati V, Iyengar R, Klotman PE, He JC: Reduction of Stat3 activity attenuates HIV-induced kidney injury. J Am Soc Nephrol 2009;20:2138–2146.

14 Eremina V, Sood M, Haigh J, Nagy A, Lajoie G, Ferrara N, Gerber HP, Kikkawa Y, Miner JH, Quaggin SE: Glomerular-specific alterations of VEGF-A expression lead to distinct congenital and acquired renal diseases. J Clin Invest 2003;111:707–716.

15 Korgaonkar SN, Feng X, Ross MD, Lu TC, D'Agati V, Iyengar R, Klotman PE, He JC: HIV-1 upregulates VEGF in podocytes. J Am Soc Nephrol 2008;19:877–883.

16 Lu TC, He JC, Wang ZH, Feng X, Fukumi-Tominaga T, Chen N, Xu J, Iyengar R, Klotman PE: HIV-1 Nef disrupts the podocyte actin cytoskeleton by interacting with diaphanous interacting protein. J Biol Chem 2008; 283:8173–8182.

17 Shkreli M, Sarin KY, Pech MF, Papeta N, Chang W, Brockman SA, Cheung P, Lee E, Kuhnert F, Olson JL, Kuo CJ, Gharavi AG, D'Agati VD, Artandi SE: Reversible cell cycle entry in adult kidney podocytes through regulated control of telomerase and Wnt signaling. Nat Med 2012;18:111–119.

18 Bruggeman LA, Ross MD, Tanji N, Cara A, Dikman S, Gordon RE, Burns GC, D'Agati VD, Winston JA, Klotman ME, Klotman PE: Renal epithelium is a previously unrecognized site of HIV-1 infection. J Am Soc Nephrol 2000;11:2079–2087.

19 Dijkman HB, Weening JJ, Smeets B, Verrijp KC, van Kuppevelt TH, Assmann KK, Steenbergen EJ, Wetzels JF: Proliferating cells in HIV and pamidronate-associated collapsing focal segmental glomerulosclerosis are parietal epithelial cells. Kidney Int 2006;70:338–344.

20 Smeets B, Uhlig S, Fuss A, Mooren F, Wetzels JF, Floege J, Moeller MJ: Tracing the origin of glomerular extracapillary lesions from parietal epithelial cells. J Am Soc Nephrol 2009; 20:2604–2615.
21 Gharavi AG, Ahmad T, Wong RD, Hooshyar R, Vaughn J, Oller S, Frankel RZ, Bruggeman LA, D'Agati VD, Klotman PE, Lifton RP: Mapping a locus for susceptibility to HIV-1-associated nephropathy to mouse chromosome 3. Proc Natl Acad Sci USA 2004;101: 2488–2493.
22 Kaufman L, Hayashi K, Ross MJ, Ross MD, Klotman PE: Sidekick-1 is upregulated in glomeruli in HIV-associated nephropathy. J Am Soc Nephrol 2004;15:1721–1730.
23 Papeta N, Chan KT, Prakash S, Martino J, Kiryluk K, Ballard D, Bruggeman LA, Frankel R, Zheng Z, Klotman PE, Zhao H, D'Agati VD, Lifton RP, Gharavi AG: Susceptibility loci for murine HIV-associated nephropathy encode trans-regulators of podocyte gene expression. J Clin Invest 2009;119:1178–1188.
24 Papeta N, Kiryluk K, Patel A, Sterken R, Kacak N, Snyder HJ, Imus PH, Mhatre AN, Lawani AK, Julian BA, Wyatt RJ, Novak J, Wyatt CM, Ross MJ, Winston JA, Klotman ME, Cohen DJ, Appel GB, D'Agati VD, Klotman PE, Gharavi AG: APOL1 variants increase risk for FSGS and HIVAN but not IgA nephropathy. J Am Soc Nephrol 2011;22:1991–1996.
25 Kaufman L, Potla U, Coleman S, Dikiy S, Hata Y, Kurihara H, He JC, D'Agati VD, Klotman PE: Upregulation of the homophilic adhesion molecule sidekick-1 in podocytes contributes to glomerulosclerosis. J Biol Chem 2010;285:25677–25685.
26 Genovese G, Tonna SJ, Knob AU, Appel GB, Katz A, Bernhardy AJ, Needham AW, Lazarus R, Pollak MR: A risk allele for focal segmental glomerulosclerosis in African-Americans is located within a region containing APOL1 and MYH9. Kidney Int 2010;78:698–704.
27 Kopp JB, Nelson GW, Sampath K, Johnson RC, Genovese G, An P, Friedman D, Briggs W, Dart R, Korbet S, Mokrzycki MH, Kimmel PL, Limou S, Ahuja TS, Berns JS, Fryc J, Simon EE, Smith MC, Trachtman H, Michel DM, Schelling JR, Vlahov D, Pollak M, Winkler CA: APOL1 Genetic variants in focal segmental glomerulosclerosis and hiv-associated nephropathy. J Am Soc Nephrol 2011; 22:2129–2137.
28 Madhavan SM, O'Toole JF, Konieczkowski M, Ganesan S, Bruggeman LA, Sedor JR: APOL1 localization in normal kidney and nondiabetic kidney disease. J Am Soc Nephrol 2011;22:2119–2128.
29 Mallipattu S, Wyatt C, He CJ: The new epidemiology of HIV-related kidney disease. J AIDS & Clin Res;201(S4):001.
30 Mallipattu SK, Liu R, Zhong Y, Chen EY, D'Agati V, Kaufman L, Ma'ayan A, Klotman PE, Chuang PY, He JC: Expression of HIV transgene aggravates kidney injury in diabetic mice. Kidney Int 2013;83:626–634.

John C. He
Department of Medicine/Nephrology, Icahn School of Medicine at Mount Sinai
One Gustave L Levy Place, Box 1243
New York, NY 10029 (USA)
E-Mail cijiang.he@mssm.edu

Liu Z-H, He JC (eds): Podocytopathy. Contrib Nephrol. Basel, Karger, 2014, vol 183, pp 191–198
DOI: 10.1159/000360114

Retinoic Acid: A Potential Pharmacologic Approach in the Treatment of Podocytopathy

Shuchita Sharma[a] · Sandeep K. Mallipattu[b] · Yifei Zhong[d] · John C. He[a, c]

[a]Division of Nephrology, Department of Medicine, Icahn School of Medicine at Mount Sinai, [b]Division of Nephrology, Department of Medicine, Stony Brook University School of Medicine, and [c]Renal Division, James J. Peters VA Medical Center, New York, N.Y., USA; [d]Department of Nephrology, Longhua Hospital, Shanghai University of Traditional Chinese Medicine, Shanghai, PR China

Abstract

In the setting of injury, podocytes can undergo dedifferentiation and proliferation, apoptosis, or cell detachment, resulting in proteinuria. Significant injury resulting in a loss of podocytes contributes to progressive kidney disease. Consequently, therapies that prevent podocyte injury and promote their regeneration will have a major clinical impact on glomerular disease. All-*trans*-retinoic acid (RA), which is a derivative of vitamin A, has many cellular functions including induction of cell differentiation, regulation of apoptosis, and inhibition of inflammation and proliferation. In addition to its established role in the treatment of variety of cancers, RA has been demonstrated to provide protection in various experimental models of kidney diseases with podocyte injury. RA produces its beneficial effects in the kidney due to its anti-inflammatory and antiproliferative effects. Also, RA helps maintain the normal morphology and preserves the differentiation markers of the podocytes. RA produces its effects in the podocytes via retinoic acid receptor-α (RAR-α). Treatment with RAR-α agonist, Am580, has a similar efficacy as RA in a mouse model of HIVAN. However, due to its systemic toxicity, clinical studies evaluating the use of RA have been challenging. Therefore, novel RAR-α agonist such as BD4 has been synthesized, which has a similar efficacy as RA and Am580, with lower toxicity.

Podocytes in Kidney Diseases

Podocytes are highly differentiated visceral epithelial cells with multiple foot processes that form one of the layers of the glomerular filtration barrier. The foot processes are attached to the glomerular basement membrane (GBM) via adhesion molecules and joined laterally by slit diaphragms. Early podocyte injury results in foot process effacement and proteinuria as seen in minimal change disease. Further podocyte damage causes proliferation, apoptosis, and loss of podocyte differentiation markers. In focal segmental glomerulosclerosis (FSGS) and diabetic nephropathy, a reduction in podocyte number results from either detachment or apoptosis, whereas collapsing FSGS of HIV-associated nephropathy (HIVAN) is characterized by proliferation and dedifferentiation of podocytes [1, 2]. Since podocytes have a limited ability to repair and regenerate, significant injury or a loss of podocytes results in development of progressive kidney disease. Thus, therapies that are aimed at preventing or limiting podocyte injury and promoting podocyte regeneration will have a major clinical impact.

Retinoic Acid

All-*trans*-retinoic acid (RA) is a signaling molecule derived from vitamin A and has multiple cellular functions which include inhibition of proliferation, induction of cell differentiation, and inhibition of inflammation [3]. Vitamin A is obtained in food and cannot be synthesized by our body. It is stored in our body in different cells, including liver and lung as inactive metabolites of vitamin A. Active metabolites, such as RA, are released when needed from specific cell types [4]. RA mediates most of its actions by binding to retinoic acid receptors (RAR) and the retinoid X receptors. RAR are expressed in many tissues, including the kidney [5]. They promote gene transcription either by binding to the RA-response elements on target genes [5, 6], or by modulating other transcription factors such as NF-κB [7] or activator protein-1 [8, 9].

Retinoic Acid in Kidney Development

RA plays a major role in nephrogenesis and induces the differentiation of kidney progenitor cells. It also affects the tubulogenesis and nephron number [10]. During kidney development, the stromal cells require RA-mediated the proximal part of the tubule, whereas uric acid is released into the gut. We will not discuss malpighian tubule in more detail here, as detailed reviews are available [5, 6].

Comparing the Fly Nephrocyte and Mammalian Podocyte

Kidney disease is the eighth leading cause of death in the United States [7]. Extensive work has been done to understand human kidney diseases using primarily cell culture systems and in vivo mammalian models. Researchers have made significant progress in understanding podocyte function, exploring the role of the slit diaphragm and basement membrane in filtration and identifying important molecules that mediate function and disease [e.g. 8–12]. Mutations in key slit-diaphragm genes can lead to severe proteinuria and kidney failure [9, 13]. Proteinuria can also be caused by defects in the reabsorption machinery in the proximal tubule, involving the major proteins in the endocytic receptor complex, e.g. Megalin, Cubilin and Amnionless (AMN) [14, 15].

Although different model systems have been used, each model has its own limitations. For example, the downregulation of some podocyte-specific proteins in cultured cells limits the utility of this system [16, 17]. Genetic manipulation in the mammalian kidney has produced important insights but is limited by time and cost. The *Drosophila* nephrocyte has clear differences with its mammalian counterpart: it is not part of a larger glomerulus but, rather, is more of a self-contained podocyte/proximal tubule. Nevertheless, the majority of molecules and cellular structures are closely conserved both in sequence and function.

Similar to mammalian podocytes, *Drosophila* pericardial nephrocytes develop from the mesoderm by the late embryo/early larval stage [18, 19]. This type of cell has been studied for more than a century, although its function has remained unclear. Kowalevsky [20] first demonstrated that cells associated with the heart, the central nervous system, and in the sternal area in scorpions can absorb ammonia carmine from the hemolymph. He suggested that those cells might function as a 'storage kidney'; the name 'nephrocyte' was introduced by Bruntz [21]. Later, Miller [22] characterized the adult *Drosophila* nephrocytes in more detail and grouped them into thoracic nephrocytes and abdominal nephrocytes. The 'abdominal nephrocytes' are now referred to as 'pericardial cells' or 'pericardial nephrocytes', as they are arranged in a row of 20–25 on each side of the heart between the first and sixth segments. The pericardial nephrocytes (approx. 40×25 μm) are larger than the thoracic nephrocytes (approx. 32×14 μm).

Liu Z-H, He JC (eds): Podocytopathy. Contrib Nephrol. Basel, Karger, 2014, vol 183, pp 224–234
DOI: 10.1159/000360119

Zebrafish Models of Podocytopathies

Rudrick C. Boucher · Weibin Zhou

Departments of Pediatrics and Communicable Diseases, University of Michigan, Ann Arbor, Mich., USA

Abstract

Over the past decade, the zebrafish *(Danio rerio)* has seen increasing use as a model system for glomerular disease. Podocyte architecture and molecular biology is highly conserved across all vertebrate clades; as such, the zebrafish is not only a highly efficient, more economical, alternative to mammalian disease models, but a versatile system for translational podocyte research by itself. Contributing to its versatility, strategies for assaying glomerular filtration and inducing podocyte injury in zebrafish have been applied to the study of podocyte gene functions. In addition, using these tools, podocyte regeneration in zebrafish has been characterized. These findings, together with other recently developed tools for zebrafish work, ensure that the zebrafish will remain a vital model system for future podocytopathy research.

Why Zebrafish?

Non-mammalian vertebrates have been extensively used as animal models throughout the history of modern kidney research [1]. Such models provide an evolutionary lens through which to refine and expand our knowledge of many facets of renal physiology and development, including the mechanisms of tubular epithelial osmoregulation, glomerular filtration, and environmental and evolutionary renal adaptation [2]. Within the scope of this review, we summarize how the zebrafish, a small tropical freshwater fish and aquatic model organism used in biomedical research, has contributed to the study of podocyte biology and emphasize the potential of this model organism as it pertains to future translational research on podocytopathies.

The zebrafish *(Danio rerio)* was established as a vertebrate genetic model organism by the pioneering work of George Streisinger and his colleagues at the

University of Oregon in the mid-1970s [3–8]. Several characteristics of the zebrafish, such as external fertilization, optical transparency of the embryo, and ease of mutagenesis and molecular manipulation, combined with its short generational time, high fecundity, and ease of care make it an ideal model system for a wide range of investigations in vertebrate genetics [9]. Building off of these inherent advantages of using the zebrafish over other model vertebrate systems, many powerful molecular, cellular and embryological techniques have been developed to further expand its versatility.

Zebrafish are easily cloned through manipulation of early embryos [8]. Cells have been traced during early embryogenesis to establish remarkably detailed cell fate maps [10, 11], and the relative ease of cell transplantation in the zebrafish facilitates genetic mosaic analysis [12, 13]. Additionally, RNA in situ hybridization and immunohistochemical labeling methods have been well tailored to investigate gene expression and protein localization in the fish [14, 15]. Direct manipulation of gene expression in zebrafish via transgenesis is also relatively simple; furthermore, microinjection of DNA or RNA allows for transient expression of genes [16]. Morpholino antisense oligonucleotides are commonly used to effectively knockdown target gene expression in zebrafish embryos [17]. Advancement of zebrafish genome data provides useful bioinformatics tools for identifying zebrafish homologs of human genes and expedites the positional cloning of causal genes for a growing collection of zebrafish mutants. Finally, several recently developed genome editing tools (e.g. TALEN [18] and CRISPR [19]) have brought reverse genetics in the zebrafish to a level equivalent to (for some applications, surpassing) that of traditional rodent models. Considering the breadth of the toolbox available, it is not surprising to see the recent explosive growth in popularity of this model organism in basic biomedical and translational research.

Podocytes and Proteinuria in Zebrafish

Vertebrate kidneys develop in three discrete stages, the pronephros, the mesonephros, and the metanephros; many vertebrate clades do not display all three forms. In most teleosts, such as the zebrafish, the pronephros is the fully functional kidney during embryonic and larval stages and the mesonephros (sometimes referred to as the 'opisthonephros' in the ichthyological literature, on account of its complexity) is the mature kidney throughout adulthood. Despite the lack of metanephros in zebrafish, the pronephric and mesonephric nephrons share striking structural and functional similarity with their metanephric counterparts in mammals and human [2].

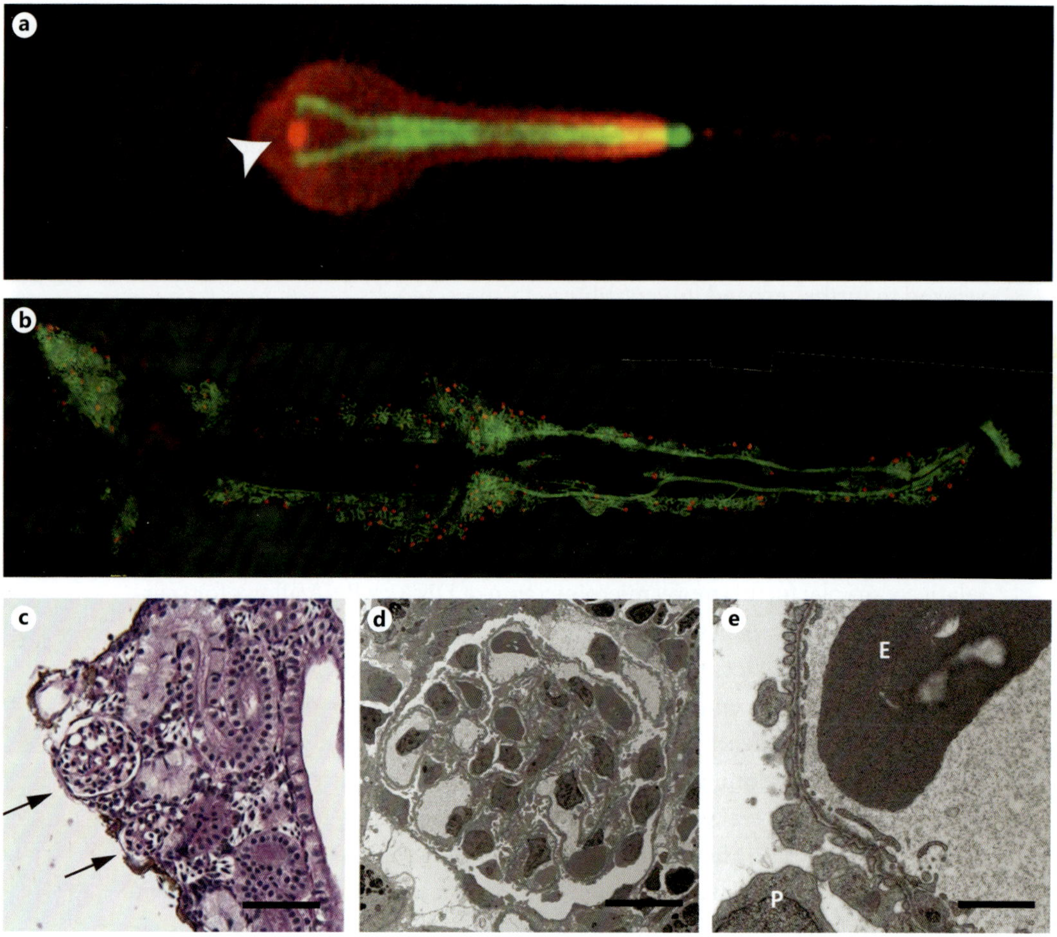

Fig. 1. Zebrafish pronephros and mesonephros. **a** Dorsal view of a 5-dpf larva of *Tg*(*nphs2*:tRFP; *enpep*:EGFP) zebrafish showing the two pronephric nephrons with the glomeruli (red, white arrowhead) fused at the midline and two tubules (green) running on each side. Anterior to the left. **b** Ventral view of the mesonephric kidney in an adult *Tg*(*nphs2*:tRFP; *enpep*:EGFP) zebrafish, with glomeruli labeled with RFP fluorescence and tubules and ducts with GFP fluorescence. Anterior to the left. **c** PAS staining of a wild-type mesonephric kidney section. Arrows: glomeruli. Scale: 50 µm. **d** Transmission electron micrograph of a mesonephric glomerulus. Scale: 10 µm. **e** Higher magnification TEM showing the ultrastructure of glomerular filtration apparatus. P = Podocyte cell body; E = erythrocyte. Scale: 1 µm.

The zebrafish pronephros comprises two nephrons, one on each side, running from the two glomeruli, which are fused at the midline, to the cloaca (fig. 1a) [1]. These pronephric nephrons are fully functioning by about 80 h post fertilization (hpf), possessing a glomerular filtration barrier (GFB) which is similar in filter selectivity to that of mammalian metanephric glomeruli [20]. At around

14 days post fertilization (dpf), the initial mesonephric nephrons begin to form. As the mesonephros takes over osmoregulation, the pronephric glomeruli degenerate by 30–60 dpf (fig. 1b) [21]. Even after the fish has reached maturity, the mesonephros continues to grow throughout the life of the fish, keeping nephron number directly proportional to body mass.

The nephron segment identity is evolutionarily conserved in zebrafish [22], although teleost nephrons lack a loop of Henle, which is an adaptation unique to mammals and birds that would presumably be superfluous in a hypo-osmotic habitat anyway [2]. Nevertheless, the structural and functional homology of renal glomeruli between zebrafish and mammals is unambiguous. Zebrafish glomeruli possess the same tissue architecture and cell populations as mammalian metanephric glomeruli. The zebrafish glomerulus contains glomerular podocytes (otherwise, this would be a very short review), fenestrated endothelial cells, mesangial cells in the glomerular tuft, and parietal epithelial cells on the wall of the Bowman's capsule. As in the metanephros, the GFB in zebrafish consists of fenestrated endothelial cells, the glomerular basement membrane and slit diaphragms formed in between the podocyte foot processes (fig. 1c). The ultrastructure of glomeruli in zebrafish, as revealed by TEM and SEM, is also highly similar to that in metanephros (fig. 1d, e).

Molecular homology between zebrafish glomeruli and metanephric glomeruli can be described from two facets. First, the molecular regulation of podocyte development is highly conserved. Familiar renal transcription factors, such as WT1, Lmx1, Foxc1/2, and Pax2 are required for podocyte development in zebrafish just as they are in the mammalian metanephros [23–25]. Second, genes essential for the function of metanephric podocytes are similarly required for podocyte function in the fish. For example, when either of two important slit-diaphragm proteins, nephrin and podocin (*nphs1* and *nphs2*, respectively), are inhibited by morpholino-mediated knockdown in zebrafish larvae, it results in severe disruption of podocyte slit-diaphram and GFB function [25].

The central role of podocytes in maintaining the normal GFB has been well established by decades of research on animal models and human nephrotic syndrome. Depletion of podocytes in rodents is sufficient to induce proteinuria and glomerular sclerosis [26] and all the known genes responsible for nephrotic syndrome are indispensable for normal podocyte function [27].

Through a transgenic zebrafish model of inducible podocyte injury, we have further demonstrated the functional conservation of podocytes in zebrafish glomeruli [28]. In this model, we used the zebrafish *nphs2* promoter to drive the expression of the *Escherichia coli* nitroreductase, enabling podocyte-specific injury with the prodrug antibiotic metronidazole. When these transgenic fish are treated with metronidazole, severe podocyte foot process effacement, podocyte loss, and apoptosis are observed, along with the disruption of GFB, proteinuria and whole-body edema. These phenotypes are highly consistent with those seen in rodent models of podocyte injury and in human nephrotic syndrome patients, further supporting the validity of zebrafish as an animal model for podocytopathies.

Proteinuria is a hallmark of various glomerular diseases in which the selectivity of the GFB is compromised. As such, methods of assessing glomerular barrier function are a prerequisite to any animal model of glomerular diseases. To this end, a number of new methods for monitoring proteinuria and GFB function in zebrafish larvae have been established. Kramer-Zucker et al. [20] initially used FITC-labeled large molecular weight dextran as a tracer of glomerular size selectivity in zebrafish. With this approach, fluorescent dextran is only detected in the pronephric ducts following histological tissue preparation of individual zebrafish larva if the GFB is compromised. Using fluorescence live imaging, Hentschel et al. [29] refined this technique into a quantitative measurement of GFB function, based on the finding that glomerular leakage of the fluorescent dextran will result in a greater reduction of FITC fluorescence in circulation in the event of GFB disruption.

Another common approach to assaying glomerular filter function is to detect the leakage of circulating serum proteins into the glomerular filtrate or excreted urine. Nishibori and his co-workers [30, 31] identified vitellogenin, a yolk protein abundant in zebrafish blood, as a major protein excreted in proteinuria when the pronephric GFB is damaged. To allow for the detection of the low concentration of excreted vitellogenin by Western blot, the total protein is concentrated by TCA precipitation from the embryo media in which a hundred zebrafish larvae are raised. This method has the advantage of providing a direct readout on GFB function using an endogenous protein, which reflects both the size and the charge selectivity of GFB.

We took advantage of the ease of zebrafish transgenesis [16] to establish a transgene-based proteinuria assay for zebrafish [28]. A transgenic zebrafish stably expressing in its serum the GFP-tagged vitamin D binding protein, which is comparable in size and charge to mammalian serum albumin, allows for highly sensitive and quantitative detection of proteinuria by ELISA against GFP, without the need to concentrate proteins. This assay is sensitive enough to detect protein leakage from a single larva or juvenile fish and can be adapted to test numerous samples simultaneously, making it potentially valuable for high-throughput screening applications.

Modeling Glomerular Diseases in Zebrafish

Ever since nephrin was identified as the first nephrotic syndrome disease gene [32] and the subsequent discovery of other podocyte-specific disease genes, rodent models of these diseases have been generated via genetic modification [33]. The ease of zebrafish genetic manipulation has made it an attractive alternative to rodent models for the functional study of disease genes and modeling of human disease. In fact, zebrafish have orthologs of all the known podocyte disease genes and zebrafish studies were a major component in functionally characterizing many of them (table 1). Transient inhibition of nephrotic syndrome gene function in zebrafish through MO-mediated knockdown causes defects in GFB function and mimics human nephrotic phenotypes with all but one *(coq6)* of the nephrotic syndrome genes thus far examined in the fish.

The zebrafish is also particularly well suited for reverse genetic studies on the functional consequences of human genetic variants, such as examining the effects of genetic variants in podocyte-specific enhancer regions in zebrafish [34]. These applications underscore the potential utility of zebrafish as a platform for the functional evaluation of genetic variants identified through high-throughput genome and exome sequencing of kidney disease patients.

Recently, zebrafish have also been used to model non-monogenic glomerulopathies, e.g. diabetic nephropathy. A number of different approaches have taken to induce chronic hyperglycemia or create a zebrafish model of type 1 diabetes. Gleeson et al. [35] treated adult zebrafish with a high concentration sucrose solution to induce hyperglycemia resulting in retinal degeneration. Intine et al. [36] used streptozocin, a toxin to pancreatic β cells, to induce β-cell deficiency, which recapitulates multiple disease phenotypes of type 1 diabetes in zebrafish, including hyperglycemia, retinopathy, defective wound healing, and pathogenic changes in the mesonephric glomeruli of zebrafish similar to early-stage diabetic nephropathy. More interestingly, a recent study on medaka fish *(Oryzias latipes)*, another common teleost model organism, has shown that high-fat diet could induce obesity-related glomerulopathy in teleosts [37]. These studies reveal the potential of using non-mammalian vertebrate models to study metabolic glomerular diseases.

Podocyte Regeneration in Zebrafish

Whereas the postnatal regenerative capacity of the mammalian kidney is extremely limited, teleosts, such as zebrafish [21] and medaka [38], retain a prolific renal regenerative capacity throughout their entire lives. Studies on meso-

Table 1. Zebrafish models of podocyte diseases

Gene name	Zebrafish phenotype	Human ortholog	Human disease	Ref.
nphs1/nephrin	Pericardiac effusion, defective GFB	*NPHS1*	SRNS	20
nphs2/podocin	Pericardiac effusion, defective GFB	*NPHS2*	SRNS	20
plce1	Pericardiac effusion, defective GFB	*PLCE1*	SRNS	43
cd2ap	Pericardiac effusion, defective GFB	*CD2AP*	SRNS	29
coq6	Massive neuronal death	*COQ6*	SRNS	44
myo1e	Pericardiac effusion, defective GFB	*MYO1E*	SRNS	30
Nostrin	Pericardiac effusion, defective GFB	*NOSTRIN*	associated with hypertension	45
mpped2	Abnormal pax2 and nephrin expression	*MPPED2*	associated with CKD	46
casp9	Reduced nephrin expression, reduced glomerular filtration clearance	*CASP9*	associated with CKD	46
Moe	Pericardiac effusion, defective GFB	*EPB41L5*		20
crb2b	Pericardiac effusion, defective GFB	*CRB2*		47
ralgps1	Pericardiac effusion, defective GFB	*RALGPS1*		47
rapgef2	Pericardiac effusion, defective GFB	*RAPGEF2*		47
cofilin-1	Pericardiac effusion, defective GFB	*CFL1*		48
myo1c	Pericardiac effusion, defective GFB	*MYO1C*		49
myh9	Reduced glomerular filtration	*MYH9*		50
lmx1b	Pericardiac effusion, pronephric glomerular cysts	*LMX1B*		24
glcci1	Collapsed glomeruli, defective GFB	*GLCCI1*		31
wt1a	Fewer differentiated podocytes	*WT1*		23
foxc1a	Fewer differentiated podocytes	*FOXC1*		23
Rbpj	Fewer differentiated podocytes	*RBPJ*		23
lat3	Collapsed glomeruli, thickened GBM, defective GFB	*LAT3*		51
p4h-tm	Fragmented GBM, defective GFB	*P4HTM*		52

GFB = Glomerular filtration barrier; SRNS = steroid-resistant nephrotic syndrome; GBM = glomerular basement membrane; CKD = chronic kidney disease.

nephric development and regeneration in zebrafish have identified the resident population of renal progenitor cells within the mesonephric mesenchyme that contributes to the ongoing nephrogenesis throughout the life cycle of the fish as well as to de novo nephron regeneration (neo-nephrogenesis) following acute renal injury [39]. However, as neither such mesenchymal renal progenitor cells nor neo-nephrogenesis have been reported in the mature metanephros of any mammalian species, the potential implications of these findings in regenerative medicine are likely to be limited. Conversely, the recent identification of podocyte precursor cells within human and rodent metanephric glomeruli [40, 41] suggests that a podocyte-regenerative approach may yet be feasible.

Intriguingly, we have found a group of epithelial cells, located at the urinary pole of zebrafish mesonephric nephron, which resembles the niche of CD133+CD24+ renal progenitor cells identified in metanephric nephrons. These cells express a teleost-specific *Wilms' tumor 1* paralog *(wt1b)*, which is a specific marker for developing nephrons and aggregating nephrogenic mesenchyme in the zebrafish mesonephric kidney [21]. Following podocyte injury, *wt1b* expression was reactivated in the parietal cells in the Bowman's capsule of mature mesonephric nephron, possibly suggesting activation of parietal cells to recapitulate podocyte ontogeny and potentially repair or replace the injured podocytes [28]. More recently, using the same metronidazole-induced podocyte injury model, Huang et al. [42] found that the pronephric podocytes could be completely regenerated, following metronidazole-mediated ablation, supporting the hypothesis that podocytes can be regenerated in the zebrafish. Since similarly activated parietal cells have been shown to contribute to both podocyte regeneration and glomerular lesions in rodents and human, the zebrafish is likely to be a valuable model for furthering the understanding of the mechanisms of these processes.

Future Outlook

Zebrafish models have only been used for podocyte work for a little over a decade. In that short time, however, they have proven themselves not only to be an indispensable adjunct to other avenues of inquiry but a versatile and valuable system in their own right. As the use of, and tools available in the, zebrafish has grown exponentially, so too has the invaluable corpus of findings made with it. In spite of evolutionary distance, the high level of conservation in glomerular architecture and podocyte biology ensure that the zebrafish will continue to act as a bridge between in vitro podocyte cultures and more traditional in vivo rodent model systems of podocyte disease.

The economy and efficiency of the zebrafish make it well suited to high-throughput functional podocyte studies which would otherwise be prohibitively difficult to perform with other vertebrate models. Studies ranging from comprehensive reverse genetic analysis of SNPs in podocyte genes, to drug screens, and even classical forward genetic screens for mutations affecting glomerular basement membrane selectivity will yield valuable results. Functional studies taking advantage of recent, and future, developments in targeted genome editing have the potential to yield a veritable treasure trove of disease models and insights on the pathogenesis of podocytopathy. Further work examining the mechanisms of podocyte regeneration is similarly promising, if only for the remote possibility that the molecular and regulatory differences between zebrafish and mammalian podocyte stem cells might inform future efforts to develop regenerative therapies for human podocytopathies.

Acknowledgements

The authors appreciate Dr. Leilei Peng for sharing the FLIM tomography images. W.Z. is a Carl W. Gottschalk Research Scholar of the American Society of Nephrology (ASN) and is supported by grants from NIH (DK091405, EB015481), the ASN Gottschalk Research Scholar Grant, the Janette Ferrantino Investigator Grant and the Young Investigator Grant from the National Kidney Foundation.

Disclosure Statement

The authors have no conflicts of interest to disclose.

References

1 Drummond IA: Kidney development and disease in the zebrafish. J Am Soc Nephrol 2005;16:299–304.
2 Dantzler WH: Comparative physiology of the vertebrate kidney. Zoophysiology, vol 22. Berlin, Springer, 1989.
3 Grunwald DJ, Streisinger G: Induction of mutations in the zebrafish with ultraviolet light. Genet Res 1992;59:93–101.
4 Grunwald DJ, Streisinger G: Induction of recessive lethal and specific locus mutations in the zebrafish with ethyl nitrosourea. Genet Res 1992;59:103–116.
5 Streisinger G, et al: Clonal origins of cells in the pigmented retina of the zebrafish eye. Dev Biol 1989;131:60–69.
6 Grunwald DJ, et al: A neural degeneration mutation that spares primary neurons in the zebrafish. Dev Biol 1988;126:115–128.
7 Streisinger G, et al: Segregation analyses and gene-centromere distances in zebrafish. Genetics 1986;112:311–319.
8 Streisinger G, et al: Production of clones of homozygous diploid zebra fish *(Brachydanio rerio)*. Nature 1981;291:293–296.

9 Detrich HW 3rd, Westerfield M, Zon LI: Overview of the zebrafish system. Methods Cell Biol 1999;59:3–10.
10 Kimmel CB, et al: Stages of embryonic development of the zebrafish. Dev Dyn 1995;203: 253–310.
11 Kohli V, Rehn K, Sumanas S: Single cell fate mapping in zebrafish. J Vis Exp 2011;pii 3172.
12 Kimmel CB, Warga RM: Indeterminate cell lineage of the zebrafish embryo. Dev Biol 1987;124:269–280.
13 Warga RM, Kimmel CB: Cell movements during epiboly and gastrulation in zebrafish. Development 1990;108:569–580.
14 Jowett T: Analysis of protein and gene expression. Methods Cell Biol 1999;59:63–85.
15 Novak AE, Ribera AB: Immunocytochemistry as a tool for zebrafish developmental neurobiology. Methods Cell Sci 2003;25:79–83.
16 Jowett T: Transgenic zebrafish. Methods Mol Biol 1999;97:461–486.
17 Nasevicius A, Ekker SC: Effective targeted gene 'knockdown' in zebrafish. Nat Genet 2000;26:216–220.
18 Sander JD, et al: Targeted gene disruption in somatic zebrafish cells using engineered TALENs. Nat Biotechnol 2011;29:697–698.
19 Hwang WY, et al: Efficient genome editing in zebrafish using a CRISPR-Cas system. Nat Biotechnol 2013;31:227–229.
20 Kramer-Zucker AG, et al: Organization of the pronephric filtration apparatus in zebrafish requires nephrin, podocin and the FERM domain protein mosaic eyes. Dev Biol 2005;285: 316–329.
21 Zhou W, et al: Characterization of mesonephric development and regeneration using transgenic zebrafish. Am J Physiol Renal Physiol 2010;299:F1040–F1047.
22 Wingert RA, Davidson AJ: The zebrafish pronephros: a model to study nephron segmentation. Kidney Int 2008;73:1120–1127.
23 O'Brien LL, et al: Wt1a, Foxc1a, and the notch mediator Rbpj physically interact and regulate the formation of podocytes in zebrafish. Dev Biol 2011;358:318–330.
24 Burghardt T, et al: LMX1B is essential for the maintenance of differentiated podocytes in adult kidneys. J Am Soc Nephrol 2013;24: 1830–1848.
25 Perner B, Englert C, Bollig F: The Wilms' tumor genes wt1a and wt1b control different steps during formation of the zebrafish pronephros. Dev Biol 2007;309:87–96.
26 Wharram BL, et al: Podocyte depletion causes glomerulosclerosis: diphtheria toxin-induced podocyte depletion in rats expressing human diphtheria toxin receptor transgene. J Am Soc Nephrol 2005;16:2941–2952.
27 Gigante M, et al: Molecular and genetic basis of inherited nephrotic syndrome. Int J Nephrol 2011;2011:792195.
28 Zhou W, Hildebrandt F: Inducible podocyte injury and proteinuria in transgenic zebrafish. J Am Soc Nephrol 2012;23:1039–1047.
29 Hentschel DM, et al: Rapid screening of glomerular slit diaphragm integrity in larval zebrafish. Am J Physiol Renal Physiol 2007; 293:F1746–F1750.
30 Mao J, et al: Myo1e impairment results in actin reorganization, podocyte dysfunction, and proteinuria in zebrafish and cultured podocytes. PLoS One 2013;8:e72750.
31 Nishibori Y, et al: Glcci1 deficiency leads to proteinuria. J Am Soc Nephrol 2011;22: 2037–2046.
32 Kestila M, et al: Positionally cloned gene for a novel glomerular protein – nephrin – is mutated in congenital nephrotic syndrome. Mol Cell 1998;1:575–582.
33 Simic I, Tabatabaeifar M, Schaefer F: Animal models of nephrotic syndrome. Pediatr Nephrol 2012;28:2079–2088.
34 He B, et al: A remote *cis*-acting variant at 3q links glomerular NCK1 to diabetic nephropathy. PLoS One 2013;8:e56414.
35 Gleeson M, Connaughton V, Arneson LS: Induction of hyperglycaemia in zebrafish *(Danio rerio)* leads to morphological changes in the retina. Acta Diabetol 2007;44:157–163.
36 Intine RV, Olsen AS, Sarras MP Jr: A zebrafish model of diabetes mellitus and metabolic memory. J Vis Exp 2013;72:e50232.
37 Ichimura K, et al: Medaka fish, *Oryzias latipes,* as a model for human obesity-related glomerulopathy. Biochem Biophys Res Commun 2013;431:712–717.
38 Watanabe N, et al: Kidney regeneration through nephron neogenesis in medaka. Dev Growth Differ 2009;51:135–143.
39 Diep CQ, et al: Identification of adult nephron progenitors capable of kidney regeneration in zebrafish. Nature 2011;470:95–100.

40 Ronconi E, et al: Regeneration of glomerular podocytes by human renal progenitors. J Am Soc Nephrol 2009;20:322–332.
41 Sagrinati C, et al: Isolation and characterization of multipotent progenitor cells from the Bowman's capsule of adult human kidneys. J Am Soc Nephrol 2006;17:2443–2456.
42 Huang J, et al: A zebrafish model of conditional targeted podocyte ablation and regeneration. Kidney Int 2013;83:1193–1200.
43 Hinkes B, et al: Positional cloning uncovers mutations in PLCE1 responsible for a nephrotic syndrome variant that may be reversible. Nat Genet 2006;38:1397–1405.
44 Heeringa SF, et al: COQ6 mutations in human patients produce nephrotic syndrome with sensorineural deafness. J Clin Invest 2011;121:2013–2024.
45 Kirsch T, et al: Knockdown of the hypertension-associated gene NOSTRIN alters glomerular barrier function in zebrafish *(Danio rerio)*. Hypertension 2013;62:726–730.
46 Pattaro C, et al: Genome-wide association and functional follow-up reveals new loci for kidney function. PLoS Genet 2012;8: e1002584.
47 Ebarasi L, et al: A reverse genetic screen in the zebrafish identifies crb2b as a regulator of the glomerular filtration barrier. Dev Biol 2009;334:1–9.
48 Ashworth S, et al: Cofilin-1 inactivation leads to proteinuria – studies in zebrafish, mice and humans. PLoS One 2010;5:e12626.
49 Arif E, et al: *Myo1c* is an unconventional myosin required for zebrafish glomerular development. Kidney Int 2013;84:1154–1165.
50 Muller T, et al: Non-muscle myosin IIA is required for the development of the zebrafish glomerulus. Kidney Int 2011;80:1055–1063.
51 Sekine Y, et al: Amino acid transporter LAT3 is required for podocyte development and function. J Am Soc Nephrol 2009;20:1586–1596.
52 Hyvarinen J, et al: Deficiency of a transmembrane prolyl 4-hydroxylase in the zebrafish leads to basement membrane defects and compromised kidney function. J Biol Chem 2010;285:42023–42032.

Weibin Zhou, PhD
8220E MSRB III, University of Michigan Medical School
1150 West Medical Center Drive
Ann Arbor, MI 48109-5646 (USA)
E-Mail weibinz@med.umich.edu

Liu Z-H, He JC (eds): Podocytopathy. Contrib Nephrol. Basel, Karger, 2014, vol 183, pp 235–244
DOI: 10.1159/000360121

The Role of MicroRNAs in Podocytes

Shaolin Shi · Zhi-Hong Liu

National Clinical Research Center of Kidney Diseases, Jinling Hospital, Nanjing University School of Medicine, Nanjing, PR China

Abstract

Background: MicroRNAs (miRNAs) are small non-coding RNA species approximately 22 nucleotides in length. MicroRNAs repress the translation of target mRNAs or degrade target mRNAs. Studies have shown that miRNAs are implicated in extensive cellular processes and are important in the pathophysiology of many cell types, including podocytes. ***Summary:*** The essential role for miRNAs in podocytes was first demonstrated by a targeted deletion of Dicer, the enzyme involved in miRNA biogenesis, in the podocytes of mice; this deletion resulted in proteinuria, podocyte injury, glomerulosclerosis, and renal failure in the mice. Further studies revealed the roles of several miRNAs in the process of podocyte injury, including dedifferentiation, adhesion disruption, and apoptosis. miRNAs also play essential roles in mesangial and endothelial cells, which are two glomerular cell types that interact with podocytes during the process of podocytopathy. Studies have also shown the roles of miRNAs in mediating the therapeutic effect of existing drugs on podocytopathies, including glucocorticoids and ACE inhibitors. Among the miRNAs in podocytes, the miR-30 family appears to be particularly important. ***Key Messages:*** miRNAs play critical roles in podocyte homeostasis and podocytopathies. Fully understanding their roles and the molecular mechanisms will improve the diagnosis and treatment of podocytopathies, but further efforts are required to achieve this understanding.

General Introduction to miRNAs

MicroRNAs are a group of small non-coding RNA molecules approximately 22 nucleotides long. A miRNA molecule can bind to a protein complex, called ribonucleoproteins (miRNPs), to form a functional RNA-induced silencing complex (RISC). The miRNA guides the RISC to the 3′-UTR of the target mRNA molecule, where a cognate sequence of this miRNA is present. As a result, the RISC represses the translation of this mRNA molecule or directly degrades it. Therefore, miRNAs function to inhibit the expression of the target genes that harbor a cognate sequence of the miRNAs. Each miRNA is predicted to target hundreds of genes [1].

There are as many as a thousand distinct miRNA genes in a mammalian genome. Some miRNA genes are located between two coding genes and are transcribed with their own promoters, and other miRNA genes are located in the intron of a coding gene and are transcribed together with the gene. After the primary transcripts of a miRNA are made, they undergo processing by Drosha, an RNase III, resulting in the release of a miRNA precursor that is approximately 70 bp in size. The miRNA precursors are transported by nuclear membrane protein exportin-5 to the cytoplasm, where the precursor is further cleaved by another RNase III, Dicer, resulting in the mature and functional form of the miRNA [2].

miRNAs are present in all types of eukaryotic cells and should maintain the expression of their target genes at optimal levels for normal cellular physiology. However, miRNA expression can be deregulated by injurious signals, resulting in alterations in the expression of their target genes. Therefore, miRNAs play important roles in many biological and pathological processes. For example, in the kidney, it was shown that miR-192 promotes TGF-β or hyperglycemia-induced collagen 1A2 expression by targeting the E-box transcription factor that acts as a transcription repressor of collagen 1A2 [3].

Requirement of miRNAs for Podocyte Homeostasis

The first evidence of the functional roles of miRNAs in podocytes was provided by studies using a mouse model with a targeted deletion of Dicer specifically in podocytes. Three groups independently reported that the targeted deletion of Dicer by the Cre-loxP system to eliminate miRNAs globally in podocytes resulted in proteinuria, loss of slit-diaphragm proteins (nephrin, podocin, and synaptopodin), podocyte foot process effacement, vacuolization, hypertrophy with crescent formation, capillary dilation, podocyte loss, glomerulosclerosis, tubulointerstitial fibrosis, and eventually renal failure and death at approximately 4–5 weeks of age [4–6]. Additionally, Ho et al. [4] profiled the glomerular miRNAs and identified the most abundantly expressed miRNAs, miR-23b, miR-24, and miR-26a. Harvey et al. [5] characterized the cytoskeletal changes in this model and determined that glomerular ezrin, moesin, and gelsolin were altered at 3 weeks of age, but the expression levels of nestin and α-actinin were unchanged. They concluded that the unique architecture of podocytes may make them particularly susceptible to cytoskeletal alterations initiated by aberrant miRNA dynamics. In the study by Shi et al. [6], gene expression profiling revealed the upregulation of 190 genes in the glomeruli of the mutant mice at the onset of proteinuria compared with the control littermates. The cognate sequences for 16 miRNAs were significantly enriched in the 3′-untranslated regions of the 190 upregulated genes, suggesting that miRNAs control target gene expression and miRNA deficiency results in upregulation of the target genes in podocytes.

Because Dicer could have functions other than miRNA synthesis, the consequence of Dicer deficiency in podocytes might not be only due to miRNA loss. To exclude this possibility, Zhdanova et al. [7] deleted Drosha, the RNAse III that is also required for miRNA biogenesis, in the mouse podocytes and observed the same phenotypes as in the Dicer mutant mice. Moreover, they used the Tet-on inducible system to delete Drosha in the podocytes of 2- to 3-month-old mice in which the podocytes were fully differentiated, and they obtained similar results. These two models have further confirmed the essential role of miRNAs in podocyte homeostasis.

In contrast to the podocytes, global miRNAs appear dispensable for the normal physiology or homeostasis of proximal tubular epithelial cells [8]. This raises a possibility that very few miRNAs among the over 100 miRNAs in podocytes are required for podocyte homeostasis and it is these miRNAs that lack Dicer-deficient podocytes which might be responsible for the observed phenotypes in the mice. This hypothesis is supported by the observation that Dicer and miR-30 knockdown in *Xenopus* had similar consequences on pronephros development [9]. Further work is needed to prove this hypothesis.

Roles of miRNAs in Podocyte Injury and Podocytopathy

A global miRNA deficiency in podocytes is known to cause podocyte and glomerular injury, but the responsible miRNAs have not yet been identified. Understanding the functions of individual miRNAs in podocytopathies represents a main challenge in the field. There are only several miRNAs that have been implicated in podocyte dedifferentiation, adhesion, and apoptosis.

Dedifferentiation

The involvement of miRNAs in the maintenance of podocyte differentiation was suggested by the observations in the mice with the targeted deletion of Dicer in the podocytes. These observations include the loss of podocyte differentiation markers (synaptopodin, nephrin, and podocin) and podocyte foot process effacement [4–6]. However, the exact miRNAs that maintain the podocyte differentiation status are unknown. Involvement of miRNAs in podocyte differentiation is also suggested by the requirement of miRNAs in the development of the pronephros in *Xenopus*. The knockdown of Dicer and Dgcr8, two key components for miRNA biogenesis, or miR-30-5p, resulted in delayed renal epithelial differentiation and abnormal patterning [9].

The best example of a miRNA that is involved in podocyte differentiation or dedifferentiation is miR-193a. Gebeshuber et al. [10] showed that the overexpression of miR-193a in mouse podocytes induced foot process effacement and eventually glomerulosclerosis. They further determined that miR-193a exerted its effect by targeting WT1. WT1 is a transcription factor that controls the differentiation and homeostasis of podocytes. It also maintains the expression of PODXL, nephrin, and several other genes that are crucial for podocyte architecture. The inhibition of WT1 by increased miR-193a triggers a catastrophic collapse of the entire podocyte-stabilizing system. Upregulation of miR-193a was observed in the glomeruli of FSGS patients.

Adhesion

Podocyte adhesion is critical for podocyte survival, and disruptions of adhesion could result in podocyte detachment and anoikis. A recent study by Li et al. [11] showed that miR-124 may regulate podocyte adhesion under the mechanical stress associated with diabetes by targeting integrin Itga3 and Itgb1.

Apoptosis

Apoptosis is considered a major cause of podocyte loss and podocytopathies [12]. The molecular mechanism underlying podocyte apoptosis is not completely understood. Recently, Long et al. [13] showed that miRNAs might be involved in podocyte apoptosis. They showed that miR-29c is upregulated in high glucose conditions, and its knockdown in vivo prevented the progression of diabetic nephropathy in db/db mice. Further mechanistic studies showed that miR-29c can induce podocyte apoptosis through targeting Sprouty homolog 1, which in turn affects Rho kinase and cause apoptosis. Chen et al. [14] found that in the diabetic nephropathy animal model DBA/2-STZ, miR-195 was upregulated in the podocytes. In vitro studies showed that the transfection of miR-195 in podocytes led to downregulation of the target BCL2, resulting in caspase-3 activation and apoptosis in podocytes.

We recently reported that the miR-30 gene family is a critical mediator in podocyte apoptosis. The miR-30 family is evolutionarily conserved and consists of 5 members, miR-30a through miR-30e. In our earlier studies, we profiled the glomerular gene expression of the mice with Dicer deficiency in podocytes and identified a group of upregulated genes. Bioinformatical analysis of these genes revealed a significant enrichment of predicted targets of miR-30s, suggesting that miR-30s profoundly regulate the gene expression in podocytes thereby functioning in podocytes [6]. Indeed, *in situ* hybridization of miR-30 with human, mouse

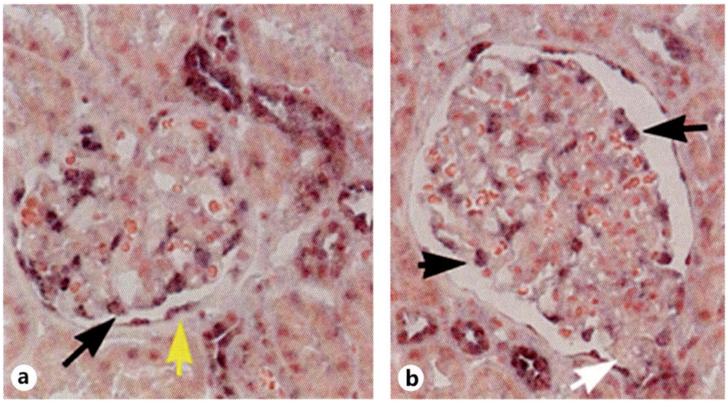

Fig. 1. a In situ hybridization shows that miR-30a is abundantly expressed in the podocytes (black arrow) and in the parietal epithelial cells (yellow arrow) but not in the mesangial or endothelial cells in normal glomeruli. **b** miR-30a is downregulated in the podocytes of FSGS patients and was confirmed by qPCR (data not shown).

and rat kidney sections showed that miR-30 members are strongly expressed in podocytes but not in mesangial and glomerular endothelial cells (fig. 1) [15]. Importantly, we have found that the miR-30 family members are all downregulated in the podocytes of FSGS patients (fig. 1) [15]. Bioinformatical analysis of the miR-30 target genes showed that apoptosis-associated genes are highly enriched in podocytes. Based on these observations, we hypothesized that miR-30 may be anti-apoptotic and that its downregulation in podocytes causes apoptosis.

We performed extensive studies to test this hypothesis and determined that the miR-30 family members can be strongly downregulated in cultured human podocytes by TGF-β, lipopolysaccharides (LPS), or puromycin aminonucleosides (PAN). The podocyte cytoskeletal damage and the apoptosis caused by TGF-β or PAN were ameliorated by exogenous miR-30 expression that sustained the cellular miR-30 levels, and the damage and apoptosis were aggravated by miR-30 knockdown. We determined that the miR-30s exert their protective roles by inhibiting the targets Notch1 and p53, which are known to mediate podocyte injury. In vivo, PAN treatment induced a strong downregulation of miR-30s in the podocytes of rats; this was accompanied by proteinuria and podocyte injury. Transferring exogenous miR-30a to the podocytes ameliorated the proteinuria, podocyte injury, and Notch signaling [15].

miRNAs Regulate the Target Genes in Podocytes That Affect Other Glomerular Cells

The vascular endothelial growth factor (VEGF) expressed and secreted by podocytes plays an important role in maintaining glomerular, particularly the endothelial cell, homeostasis, and its expression alteration occurs in some glomerular diseases including diabetic nephropathy [16]. The mechanisms by which VEGF expression is regulated are not understood. Long et al. [17] recently showed that miR-93 regulates VEGF-A expression in the podocytes of both in vitro and in vivo diabetic nephropathy models, which implicates miR-93 in endothelial cell injury by inhibiting VEGF expression in podocytes.

Some insight into the roles of miRNAs in podocytopathies have been provided by different studies, however a large amount of future work remains because there are more than 100 miRNAs that are significantly expressed in podocytes [pers. unpubl. data].

Roles of miRNAs in Glomerular Mesangial or Endothelial Cells

It is believed that three types of glomerular cells, podocytes, mesangial and endothelial cells, interact with each other and facilitate the progression of glomerular diseases [18]. Therefore, the deregulation of miRNAs and the resultant alterations of target gene expression in these cells could affect the podocytes.

There have been quite a few studies investigating the roles of miRNAs in mesangial cells. Several studies have shown that miR-192, miR-29 and miR-216 control the expression of collagens in the mesangial cells and regulate the ECM accumulation and renal fibrosis [3, 19, 20]. miR-377 and miR-29c have been shown to regulate fibronectin expression [13, 21]. miRNAs are also implicated in other cellular processes in mesangial cells. Altered expression levels of miR-216 and miR-217 lead to abnormal survival and hypertrophy of the mesangial cells [22]. miR-21 inhibits the proliferation of the mesangial cells [23], and miR-29c promotes mesangial cell apoptosis [13].

Relatively few studies have been performed to investigate the roles of miRNAs in glomerular endothelial cells. Asgeirsdóttir et al. [24] showed that miR-126 has a major role in the segmental, heterogenic response of the renal microvascular endothelial cells to systemic inflammatory stimuli by controlling the expression of VCAM-1, an endothelial adhesion molecule.

Integration of miRNAs into Podocyte Signaling Networks

The rapidly expanding knowledge regarding gene expression regulation by miRNAs needs to be integrated into the existing signaling network that is based on transcriptomics, proteomics, phosphoproteomics, metabolomics, and epigenomics to obtain a full picture of the regulatory network. Such attempts have been made, for example miRNAs have been incorporated into the regulatory network of general cells in the bioinformatics analysis tool Ingenuity (www.ingenuity.com). It is also important to generate cell or tissue type specific regulatory networks [25]. For podocytes, Boerries et al. [26] performed mRNA, miRNA, and quantitative proteomic analyses on the podocytes isolated from mice; they revealed programs of highly specific gene regulation tightly controlling the cytoskeleton, cell differentiation, endosomal transport, and peroxisome function in podocytes by integrative analyses of these different 'omics' data. This integrative 'omics' approach will facilitate podocytopathy research.

miRNAs as Therapeutic Targets for Podocytopathies

Because of their critical roles in podocyte pathophysiology, miRNAs can potentially serve as drug targets for podocytopathy. Two studies showed that miRNAs mediate the effects of the existing drugs for podocytopathies. Using the nephropathy model of Munich Wistar Frömter rats, Macconi et al. [27] showed that upregulation of miR-324-3p in podocytes and tubular cells promoted renal fibrosis by inhibiting the target prolyl endopeptidase (Prep), which is a serine peptidase involved in the metabolism of angiotensins and the synthesis of the antifibrotic peptide N-acetyl-seryl-aspartyl-lysyl-proline (Ac-SDKP). The authors further showed that ACE inhibitors reversed the process by downregulating miR-423-3p. Therefore, the renoprotective effects of ACE inhibitors may be partially a result of miR-324-3p modulation.

In our recent study, we showed that glucocorticoids could prevent the downregulation of miR-30s by TGF-β, LPS or PAN in cultured podocytes. Glucocorticoid-treated rats exhibited a less significant miR-30 reduction in podocytes, which was accompanied by reduced proteinuria and podocyte injury, with PAN treatment [15]. Figure 2 is a model demonstrating the relationships among injurious factors (e.g. TGF-β), miR-30s, Notch or p53, and glucocorticoids in the podocytes. Because glucocorticoids have been shown to have a direct protective effect on the cytoskeleton and to prevent podocyte apoptosis [28, 29], our studies provided an additional mechanism underlying the protective effect of glucocorticoids on podocytes.

Conclusions

Recent work has clearly demonstrated the essential roles for miRNAs in podocyte development, homeostasis and injury. Studies have also shown the involvement of miRNAs in podocytopathy treatment. miRNAs exert their effects in

Fig. 2. A model depicting the interactions between the factors involved in podocyte injury. Injurious factors activate Notch1 and p53 signaling by downregulating miR-30s, resulting in podocyte apoptosis and podocytopathy. Glucocorticoids prevent miR-30 downregulation, which prevents Notch1 and p53 activation, podocyte apoptosis, and podocytopathy.

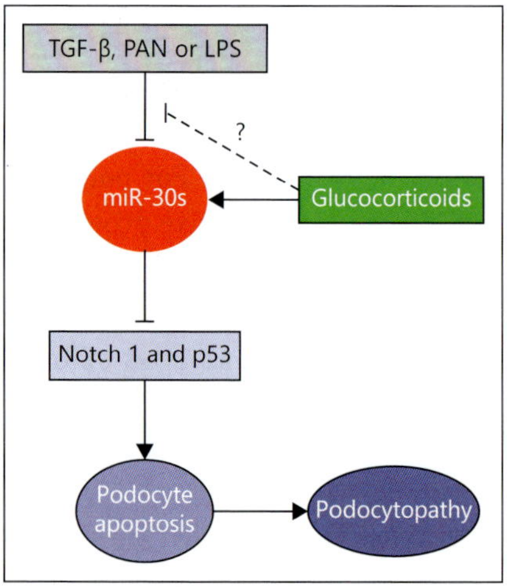

podocytes mainly through regulating the expression of the genes that play a critical role in podocyte homeostasis and the development of podocytopathy. Fully understanding their functional roles and the molecular mechanisms will certainly help improve the diagnosis and treatment of podocytopathies.

Acknowledgements

Z.-H.L. and S.S. were supported by the National Basic Research Program of China 973 Program No. 2012CB517600 (No. 2012CB517606), the Key Projects of the Natural Science Foundation of Jiangsu Province (BK2011020, BK20131324), and the National Natural Science Foundation of China (81370827).

Disclosure Statement

The authors have no conflicts of interest to disclose.

References

1 John B, Enright AJ, Aravin A, Tuschl T, Sander C, Marks DS: Human MicroRNA targets. PLoS Biol 2004;2:e363.
2 Murchison EP, Hannon GJ: miRNAs on the move: miRNA biogenesis and the RNAi machinery. Curr Opin Cell Biol 2004;16:223–229.
3 Kato M, Zhang J, Wang M, Lanting L, Yuan H, Rossi JJ, Natarajan R: MicroRNA-192 in diabetic kidney glomeruli and its function in TGF-β-induced collagen expression via inhibition of E-box repressors. Proc Natl Acad Sci USA 2007;104:3432–3437.

4 Ho J, Ng KH, Rosen S, Dostal A, Gregory RI, Kreidberg JA: Podocyte-specific loss of functional microRNAs leads to rapid glomerular and tubular injury. J Am Soc Nephrol 2008; 19:2069–2075.
5 Harvey SJ, Jarad G, Cunningham J, Goldberg S, Schermer B, Harfe BD, McManus MT, Benzing T, Miner JH: Podocyte-specific deletion of Dicer alters cytoskeletal dynamics and causes glomerular disease. J Am Soc Nephrol 2008;19:2150–2158.
6 Shi S, Yu L, Chiu C, Sun Y, Chen J, Khitrov G, Merkenschlager M, Holzman LB, Zhang W, Mundel P, Bottinger EP: Podocyte-selective deletion of Dicer induces proteinuria and glomerulosclerosis. J Am Soc Nephrol 2008; 19:2159–2169.
7 Zhdanova O, Srivastava S, Di L, Li Z, Tchelebi L, Dworkin S, Johnstone DB, Zavadil J, Chong MM, Littman DR, Holzman LB, Barisoni L, Skolnik EY: The inducible deletion of Drosha and microRNAs in mature podocytes results in a collapsing glomerulopathy. Kidney Int 2011;80:719–730.
8 Wei Q, Bhatt K, He HZ, Mi QS, Haase VH, Dong Z: Targeted deletion of Dicer from proximal tubules protects against renal ischemia-reperfusion injury. J Am Soc Nephrol 2010;21:756–761.
9 Agrawal R, Tran U, Wessely O: The miR-30 miRNA family regulates *Xenopus* pronephros development and targets the transcription factor Xlim1/Lhx1. Development 2009;136: 3927–3936.
10 Gebeshuber CA, Kornauth C, Dong L, Sierig R, Seibler J, Reiss M, Tauber S, Bilban M, Wang S, Kain R, Böhmig GA, Moeller MJ, Gröne HJ, Englert C, Martinez J, Kerjaschki D: Focal segmental glomerulosclerosis is induced by microRNA-193a and its downregulation of WT1. Nat Med 2013;19:481–487.
11 Li D, Lu Z, Jia J, Zheng Z, Lin S: Changes in microRNAs associated with podocytic adhesion damage under mechanical stress. J Renin Angiotensin Aldosterone Syst 2013;14:97–102.
12 Schiffer M, Bitzer M, Roberts IS, Kopp JB, ten Dijke P, Mundel P, Böttinger EP: Apoptosis in podocytes induced by TGF-β and Smad7. J Clin Invest 2001;108:807–816.
13 Long J, Wang Y, Wang W, Chang BH, Danesh FR: MicroRNA-29c is a signature microRNA under high glucose conditions that targets Sprouty homolog 1, and its in vivo knockdown prevents progression of diabetic nephropathy. J Biol Chem 2011;286: 11837–11848.
14 Chen YQ, Wang XX, Yao XM, Zhang DL, Yang XF, Tian SF, Wang NS: MicroRNA-195 promotes apoptosis in mouse podocytes via enhanced caspase activity driven by BCL2 insufficiency. Am J Nephrol 2011;34:549–559.
15 Wu J, Zheng CX, Fan Y, Zeng CH, Chen ZH, Qin WS, Zhang CM, Zhang WF, Wang X, Zhu XD, Zhang MC, Zen K, Liu ZH: Downregulation of microRNA-30 facilitates podocyte injury and can be prevented by glucocorticoids. J Am Soc Nephrol 2014;25:92–104.
16 Sison K, Eremina V, Baelde H, Min W, Hirashima M, Fantus IG, Quaggin SE: Glomerular structure and function require paracrine, not autocrine, VEGF-VEGFR-2 signaling. J Am Soc Nephrol 2010;21:1691–1701.
17 Long J, Wang Y, Wang W, Chang BH, Danesh FR: Identification of microRNA-93 as a novel regulator of vascular endothelial growth factor in hyperglycemic conditions. J Biol Chem 2010;285:23457–23465.
18 Schlöndorff D, Banas B: The mesangial cell revisited: no cell is an island. J Am Soc Nephrol 2009;20:1179–1187.
19 Wang B, Komers R, Carew R, Winbanks CE, Xu B, Herman-Edelstein M, Koh P, Thomas M, Jandeleit-Dahm K, Gregorevic P, Cooper ME, Kantharidis P: Suppression of microRNA-29 expression by TGF-β_1 promotes collagen expression and renal fibrosis. J Am Soc Nephrol 2012;23:252–265.
20 Kato M, Wang L, Putta S, Wang M, Yuan H, Sun G, Lanting L, Todorov I, Rossi JJ, Natarajan R: Post-transcriptional up-regulation of Tsc-22 by Ybx1, a target of miR-216a, mediates TGF-β-induced collagen expression in kidney cells. J Biol Chem 2010;285:34004–34015.
21 Wang Q, Wang Y, Minto AW, Wang J, Shi Q, Li X, Quigg RJ: MicroRNA-377 is up-regulated and can lead to increased fibronectin production in diabetic nephropathy. FASEB J 2008;22:4126–4135.

22 Kato M, Putta S, Wang M, Yuan H, Lanting L, Nair I, Gunn A, Nakagawa Y, Shimano H, Todorov I, Rossi JJ, Natarajan R: TGF-β activates Akt kinase through a microRNA-dependent amplifying circuit targeting PTEN. Nat Cell Biol 2009;11:881–889.
23 Zhang Z, Peng H, Chen J, Chen X, Han F, Xu X, He X, Yan N: MicroRNA-21 protects from mesangial cell proliferation induced by diabetic nephropathy in db/db mice. FEBS Lett 2009;583:2009–2014.
24 Asgeirsdóttir SA, van Solingen C, Kurniati NF, Zwiers PJ, Heeringa P, van Meurs M, Satchell SC, Saleem MA, Mathieson PW, Banas B, Kamps JA, Rabelink TJ, van Zonneveld AJ, Molema G: MicroRNA-126 contributes to renal microvascular heterogeneity of VCAM-1 protein expression in acute inflammation. Am J Physiol Renal Physiol 2012;302:F1630–F1639.
25 Velic A, Macek B, Wagner CA: Toward quantitative proteomics of organ substructures: implications for renal physiology. Semin Nephrol 2010;30:487–499.
26 Boerries M, Grahammer F, Eiselein S, Buck M, Meyer C, Goedel M, Bechtel W, Zschiedrich S, Pfeifer D, Laloë D, Arrondel C, Gonçalves S, Krüger M, Harvey SJ, Busch H, Dengjel J, Huber TB: Molecular fingerprinting of the podocyte reveals novel gene and protein regulatory networks. Kidney Int 2013;83:1052–1064.
27 Macconi D, Tomasoni S, Romagnani P, Trionfini P, Sangalli F, Mazzinghi B, Rizzo P, Lazzeri E, Abbate M, Remuzzi G, Benigni A: MicroRNA-324-3p promotes renal fibrosis and is a target of ACE inhibition. J Am Soc Nephrol 2012;23:1496–1505.
28 Faul C, Donnelly M, Merscher-Gomez S, Chang YH, Franz S, Delfgaauw J, Chang JM, Choi HY, Campbell KN, Kim K, Reiser J, Mundel P: The actin cytoskeleton of kidney podocytes is a direct target of the antiproteinuric effect of cyclosporine A. Nat Med 2008; 14:931–938.
29 Wada T, Pippin JW, Marshall CB, Griffin SV, Shankland SJ: Dexamethasone prevents podocyte apoptosis induced by puromycin aminonucleoside: role of p53 and Bcl-2-related family proteins. J Am Soc Nephrol 2005; 16:2615–2625.

Zhi-Hong Liu, MD
National Clinical Research Center of Kidney Diseases
Jinling Hospital, Nanjing University School of Medicine
305 East Zhongshan Road, Nanjing 210002 (PR China)
E-Mail liuzhihong@nju.edu.cn

Liu Z-H, He JC (eds): Podocytopathy. Contrib Nephrol. Basel, Karger, 2014, vol 183, pp 245–255
DOI: 10.1159/000360123

An Integrative Biology Approach to Study Podocyte Function

Wenjun Ju[a, b] · Jeffery B. Hodgin[c] · Song Jiang[d] · Matthias Kretzler[a, b]

Departments of [a]Internal Medicine, [b]Computational Medicine and Bioinformatics, and [c]Pathology, University of Michigan, Ann Arbor, Mich., USA; [d]National Research Center of Kidney Diseases, Jinling Hospital, Nanjing University School of Medicine, Nanjing, China

Abstract

The podocyte plays important role in maintaining an intact filtration barrier, and defect or loss of podocytes is a major cause of inherited and acquired glomerular diseases. Efforts towards a comprehensive understanding of the role of the podocyte in the glomerular pathobiology leading to novel ways to target the podocyte for therapeutic intervention remain a major research focus in nephrology. The development of molecular medicine allows us to generate comprehensive profiles of information at genomic, transcriptomic, proteomic and metabolomic levels to define podocyte function and failure. Integrating data from two or more of these approaches improves functional annotations and facilitates the formulation of biological hypotheses. We will review the definition of systems biology, the integration between large-scale datasets, the current status of the systems biology approach in podocyte research, as well as the challenges and opportunities in the field. We also will introduce relevant databases and useful tool suites that can help to efficiently extract the relevant information out of the sometime overwhelming amounts of data generated. Integrating the emerging large-scale datasets in podocyte research will advance our understanding of podocyte development and function, pathophysiology of glomerular diseases, and identification of podocyte-centric therapeutic targets.

Systems Biology Is Integrative Biology

An integrative or systems biology approach incorporates large-scale datasets from diverse sources to achieve a more holistic understanding of health and disease. The goal of systems biology as described in this review is to integrate interconnected genetic variances, transcriptomic profiles, protein expression, and

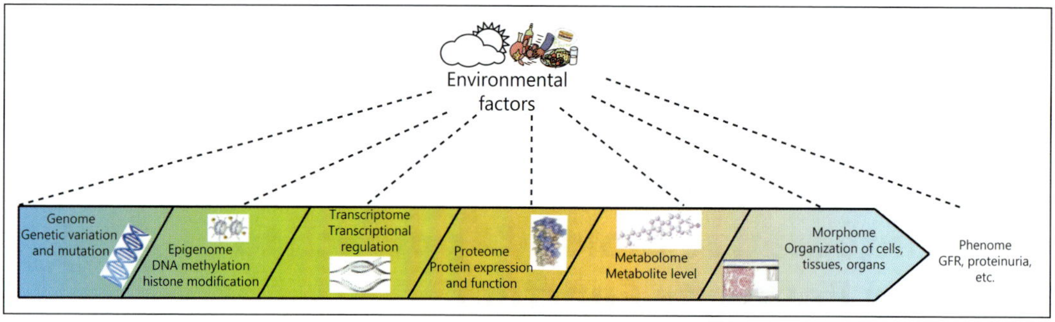

Fig. 1. An integrative biology view of large-scale datasets along the genome-phenome continuum (figure is adapted from a review article by Kretzler and Cohen [53]). Technologies applied to generate large-scale datasets have been previously described [4]. In brief, genome sequencing, high-density SNP chips, methylation assays and histone modification analysis for genome/epigenome (genetic variation, mutation, DNA methylation and histone modification) analysis; microarray and RNA sequencing for transcriptional regulation; yeast two-hybrid, co-immunoprecipitation and 2D differential gel electrophoresis for proteome (protein expression and function) analysis; nuclear magnetic resonance spectroscopy and mass spectrometry for metabolome (sugars, lipids, amino acids, small molecules) analysis; histology, immunohistochemistry and electron microscopy for morphome (organization of cells, tissues and organs) analysis; clinical chemistry, clearance measurement, imaging studies for phenome (glomerular filtration rate (GFR), proteinuria level, etc.) analysis.

metabolite levels to generate novel pathophysiological perspectives. Applying a systems biology approach to study kidney disease has been comprehensively discussed and omics datasets derived or associated with kidney disease have been thoroughly reviewed by He et al. [1] in their recent review. Here we will not review the individual experimental strategies deployed by each of these techniques, but rather focus on the importance of integration of large-scale datasets to define underlying key biological principles. In addition, omics data generated from urinary sediment are not in the scope of this review.

Integration can occur across every level of the genome-phenome continuum (fig. 1) to achieve cross-cutting information beyond that of each individual approach.

The approach is currently employed for the integration of genomic and transcriptomic data, which has been termed systems genetics [2]. The translation of common genetic variants identified through genome-wide association studies (GWAS) for the risk and manifestation of common kidney diseases into mechanistic disease insight is a major challenge in the field. These studies have been very successful in identifying a statistical significant association of a renal phenotype (= trait), but do not provide knowledge or analysis of mechanisms and intermediate regulatory cascades linking the trait to the genetic risk locus. In addition, the discovered variants frequently have a small effect size, which makes the association between disease and variant challenging to evaluate experimentally. To bridge the gap between genetic variants and clinical phenotype traits in nephrology, Keller et al. [2] provided a plausible strategy integrating genetic risk loci with kidney-specific, genome-wide gene expression. Risk loci-associated transcripts can then be assigned a putative disease-specific function using gene expression co-regulation networks. Integration of genotypic information with intermediary molecular traits, such as gene expression pattern or metabolites profile, is defined as expression quantitative trait locus (eQTL) or metabolic quantitative trait locus (mQTL) [3, 4]. This approach has been proven to be helpful in delineating the molecular impact of the risk loci on the outcome [5], however matching a GWAS SNP to an eQTL remains a challenging task [2, 6, 7].

An integrated transcriptomic/proteomic and metabolomic analysis at the cellular or tissue level allows mapping of the differentially regulated genes/proteins and metabolites into the same metabolic pathway. This offers added confidence in identifying the most promising pathways involved in phenotypic change or pathogenesis of the disease of interest. One example is integration of transcriptomic and metabolomic analysis on the primary cultured inner medullary collecting duct cells. Choi et al. [8] revealed that decreased extracellular osmolality is associated with decreased levels of organic osmolytes, glucose, intermediates of citric acid cycle, and branched-chain amino acids, along with significantly decreased gene expression and protein abundance of certain transporters and insulin-signaling pathways [8].

Current Status of Integrative Biology of Podocyte

The majority of the kidney omics data has been derived from whole kidney tissue. Transcriptome analyses have also been applied to normal and diseased glomeruli microdissected from mice [9–11], rats [12], and humans [13–15], revealing the potential for glomerular transcriptomes to identify pathogenic pathways [12]. Proteomic profiles have been reported as well for healthy and diseased glomeruli from mice [16–19], rats [20, 21] and humans [22–27]. However, despite the clear importance of podocytes in glomerular disease, as reviewed in previous chapters, studies focusing on podocytes are still at very early stage. This is mainly due to technical difficulties which will be described in detail in the following paragraphs. As of today, the published omics studies on podocytes include transcriptomic analysis of mouse podocytes [28–31] and proteomics analysis of mouse [28, 32] and human [33] podocytes.

Pioneering work from Boerries et al. [28] reported a comprehensive characterization of the transcriptome and proteome of endogenous mouse podocytes.

Mouse podocytes genetically engineered to express green fluorescent protein were isolated by fluorescent-activated cell sorting (FACS) to allow best representation of in vivo podocytes, in contrast to established in vitro cell lines. This study not only provided rich podocyte-enriched transcriptomic and proteomic information, but also integrated functional analyses using combined datasets. The analysis revealed 7 network clusters based on gene ontology-term groups overrepresented in podocytes as starting points for further analyses: endoplasmic reticulum, ubiquitination, cytoskeleton, nuclear elements and mitochondria, peroxisome and protein transport, and cell junction.

Brunskill et al. [31] applied a similar approach to facilitate the FACS-based purification of podocytes from murine metanephric development stage at E13.5 and E15.5, as well as adults. Their transcriptome analysis identified multiple sets of genes that establish the neuronal, muscle, and phagocytic properties of podocytes, characterized combinatorial code of transcription factors, and defined the global lists of growth factors and receptors that are expressed by podocytes.

In a unique approach, Warsow et al. [34] built a literature-based mouse podocyte network, PodNet, and used database protein-protein interactions (PPI) to expand PodNet to XPodNet. By mapping and analyzing transcriptomes of podocytes in 10 different states on XPodNet, they found that interactions between slit diaphragm proteins are mostly upregulated during podocyte development, but mostly downregulated in cell culture. In addition, their analysis indicated that interactions lost during developmental podocyte differentiation are not regained in cell culture. These podocyte PPI networks and differential interactome analyses can serve as a powerful resource for podocyte research.

The work from these laboratories provides an atlas of gene and protein expression patterns of developing and adult podocytes that can guide further, focused podocyte research. Despite this early progress in podocyte transcriptome and proteome study, further studies are required to complement these genome scale datasets. Upcoming epigenetic and metabolomic data of podocytes will undoubtedly provide crucial information (e.g. how environmental factors may impact cellular behavior through epigenetic and metabolomic regulation) in comprehensive understanding of podocyte biology.

Novel Technical Breakthroughs Facilitate Integrative Biology Study of Podocytes

The key challenge to study podocytes is the technical difficulty to obtain a sufficient number of podocytes, with minimal contamination of the neighboring mesangial, endothelial or parietal epithelial cells, that resemble the functionality of in vivo podocytes. Previous high-throughput strategies have relied on mouse [35, 36] or human [37] immortalized cells which have well-known drawbacks, such as rapid loss of podocyte-specific phenotypes and marker gene expression profile [38, 39] and inconsistent divergent phenotypes that are likely caused by various culture conditions [28]. In contrast to established in vitro cell lines, FACS of genetically engineered mouse podocytes [28, 31], as mentioned in the previous paragraph, improves the representation of in vivo podocytes and provides advantages in comparison with in vitro cell lines.

Cell sorting via fluorescence has not been used to isolate human podocytes for molecular analysis. In addition to human immortalized podocyte cell lines, current studies on human podocytes have relied on the findings from murine podocytes based on the assumption that human podocytes share genes that perform similar functions across species. This assumption does not hold for several molecular pathways with significant functional implications in human disease. The most recent example relevant to the podocyte is the apolipoprotein L1 gene *(APOL1)*, whose genetic variants are well known to be associated with focal segmental glomerulosclerosis [40–42], HIV-associated nephropathy [40], and hypertensive nephropathy [43] in individuals of African descent. *APOL1* is expressed in human podocytes, but interestingly the mouse does not express an *APOL1* ortholog [44]. In fact, despite significant similarities shared by human and mouse genome, only 75% of mouse and 80% of human genes are in a simple one-to-one orthologous relationship [45]. Of those one-to-one orthologous genes, over 20% of human essential genes have non-essential mouse orthologs [46], suggesting these genes carry distinct functionality in human and mice.

The genomic differences across species argue that we need alternative methods to identify human podocyte-specific genes. Recent work from our group has tackled this barrier using a computational approach [47], in silico nanodissection. This machine learning-based approach leverages high-throughput RNA expression data derived from microdissected human kidney biopsy tissue. Using a novel iterative statistical framework, we identified transcripts specifically enriched in podocytes on a genome scale, and obtained predictions more accurate than those based on cross-species comparison, when evaluated by an independent resource, i.e. the Human Protein Atlas (HPA) [48]. Notably, *APOL1* is among the genes from our computational approach with a high predictive score. This study highlights the importance of integrating experimental approaches and computational biology to convert information into novel knowledge. The approach allows us to take advantage of accumulated omics data derived from diseased kidney tissue/glomeruli, and by adding a filter of cell lineage specificity, deconvolute the functionality at cell type level (podocytes, mesangial cells, and endothelial cells). The lineage-specific information is useful not only for inves-

Table 1. Databases that may facilitate podocyte research (with open access)

Data type	Database	Link
General database and tool suite	NCBI	http://www.ncbi.nlm.nih.gov/
Genome database	Ensembl	http://useast.ensembl.org/index.html
Protein/RNA expression	GUDMAP	http://www.gudmap.org/
Protein/RNA expression	HPA	http://www.proteinatlas.org/
RNA expression	Nephromine	http://www.nephromine.org/

tigation of pathomechanism, but also allows developing targeted interventional approaches. In addition, by providing a systematic scoring system for human podocyte enrichment, we can help to prioritize the putative causal genes identified by genetic studies across hereditary diseases that are often caused by cell lineage-specific transcripts [49]. Vice versa, identified podocyte-enriched genes (at http://nano.princeton.edu/) can be screened comprehensively in families with a hereditary glomerular disease.

Database and Freely Accessible Bioinformatics Tool Suites to Facilitate Systems Biology Research on Podocytes

An integrative biology approach requires access to a well-established database as well as up-to-date and user-friendly computational tools. Many publically available databases have been established and continuously updated that allow researchers to mine kidney relevant data (table 1). The most general and widely used is National Center for Biotechnology Information (NCBI). NCBI provides comprehensive databases and tools for biomedical research including databases of genomic structural variation (dbVar) (archives information associated with large-scale genomic variation), genome (harbors sequence and map data from the whole genomes of over 1,000 organisms), epigenetics (contains genome-wide maps of DNA and histone modifications from a wide collection of epigenetic datasets), nucleotide database (collects nucleotide sequences from several sources, including GenBank, RefSeq, the third-party annotation database, and protein databank), and much more. The Ensembl is another very widely used genome databases for vertebrates and other eukaryotic species. Additional databases we have found especially useful include Genitourinary Development Molecular Anatomy Project (GUDMAP), the HPA, and Nephromine (table 1).

GUDMAP is an NIH-funded consortium of laboratories working to provide the scientific and medical community with tools to facilitate kidney-related research [50]. The most recently added key components are high-resolution RNA expression data at specific cell type level including glomerular podocytes, mesangial cells, and glomerular capillary endothelial cells [28, 51, 52]. It provides a valuable resource to research on murine podocyte gene expression signature.

The HPA [48] provides high-resolution immunohistochemistry (IHC) or immunofluorescence images for proteins of more than 15,156 genes based on 18,707 antibodies (data per HPA 11.0-2013.03.11) in 46 different normal human tissues, including kidney. In the HPA, the IHC staining for some proteins on kidney tissue exhibits high-resolution allowing detection the expression pattern of podocyte genes [47, 48] through visualizing the IHC staining images by experienced researchers.

Finally, Nephromine [53] is a web-based renal centric gene expression data repository, search and analytical tool with integrated systems biology tools, allowing users to extract key information to target ongoing kidney disease and to generate de novo hypotheses for further experimental investigations. Nephromine does not directly provide transcriptome information at cell type level, however, users can upload a list of podocyte-enriched genes or genes that are potentially associated with podocyte-specific structure or function, as a customized concept, to query Nephromine for kidney disease-related regulation patterns for those podocyte-associated genes. The current disease etiologies included in Nephromine are: focal segmental glomerulosclerosis, minimal change disease, diabetic nephropathy, hypertension, IgA nephropathy, lupus nephritis, etc.

In addition to the above databases, user-friendly computational tool suites are also freely available. Table 2 is a non-exhaustive list of those tools we have found helpful to extract maximal information out of the large amounts of data generated by systems biology analysis.

Summary and Perspectives

Genetically engineered murine podocytes enabled highly improved purification of in vivo murine podocytes, and the in silico nanodissection approach allowed prediction of novel human podocyte-enriched genes. The technologies allowing to apply an integrative or systems biology approach to human podocytes are currently emerging. These technical advances may help to accelerate podocyte research at genomic, epigenetic, transcriptomic, proteomic, metabolomic level, to obtain a comprehensive view of podocyte cellular process in health and disease. Major challenges, some of which have been discussed above, need to be taken on include data integration (such as integration between various omics data, integration between omics data and cellular responses, morphometric data and

Table 2. Computational tool suites that may facilitate integrative biology research (with open access)

Application	Program	Link
PPI, Concepts, Metscape	NCIBI suite	http://portal.ncibi.org/gateway/tryourtools.html
PPI	STRING	http://string-db.org/
Analysis platform and data processing	GenePattern	http://www.broadinstitute.org/cancer/software/genepattern/
Analysis platform and data processing	Bioconductor	http://www.bioconductor.org/
Array analysis and visualization	MultiExperiment viewer	http://www.tm4.org/mev.html
Array analysis and visualization	ArrayTrack	http://edkb.fda.gov/webstart/arraytrack/
Functional annotation and enrichment analysis	DAVID	http://david.abcc.ncifcrf.gov/
Enrichment analysis	GSEA	http://www.broadinstitute.org/gsea/index.jsp
Enrichment and candidate gene prioritization	ToppGene Suite	http://toppgene.cchmc.org/

NCIBI = The National Center for Integrative Biomedical Informatics; STRING = Search Tool for the Retrieval of Interacting Genes/Proteins; DAVID = The Database for Annotation, Visualization and Integrated Discovery; GSEA = Gene Set Enrichment Analysis; PPI = protein-protein interaction.

patients' clinical parameters) and how to transform the descriptive data to mechanistic knowledge of podocyte function in mouse and man. Most importantly, the definition of the mechanistic context of podocyte failure is critical to help clinical decision-making via a functional disease definition leading to improved diagnosis and prognosis. Ultimately, how can these data direct us toward targeted therapeutic strategies, which may require integration of these data with drug and kidney disease databases. A network of clinical translational research studies deploying harmonized clinical and molecular study protocols is currently emerging among sites in Asia, the USA, Europe and Africa [for an example of a study protocol, see 54, 55]. These networks can serve as platforms to study podocytopathies in humans using a systems biology approach, where molecular profiles will be linked to phenotypic, genetic, and digitalized histological data for comprehensive analyses and understanding the complexity of the disease pathogenesis. The hope is that these cohorts will be in place to translate the mechanistic insights from model systems into care of patients in targeted interventional trials.

Acknowledgements

The authors thank Dr. Markus Bitzer for insightful discussion and Dr. Celine Berthier for critical help with adaption of the figure. This study was supported in part by R01 GM071966, R01 DK079912 and P30 DK081943 to M.K.

Disclosure Statement

The authors have no conflicts of interest to disclose.

References

1 He JC, Chuang PY, Ma'ayan A, Iyengar R: Systems biology of kidney diseases. Kidney Int 2012;81:22–39.
2 Keller BJ, Martini S, Sedor JR, Kretzler M: A systems view of genetics in chronic kidney disease. Kidney Int 2012;81:14–21.
3 Keller B, Martini S, Sedor J, Kretzler M: Linking variants from genome-wide association analysis to function via transcriptional network analysis. Semin Nephrol 2010;30:177–184.
4 Komorowsky CV, Brosius FC 3rd, Pennathur S, Kretzler M: Perspectives on systems biology applications in diabetic kidney disease. J Cardiovasc Transl Res 2012;5:491–508.
5 Wheeler HE, et al: Sequential use of transcriptional profiling, expression quantitative trait mapping, and gene association implicates MMP20 in human kidney aging. PLoS Genet 2009;5:e1000685.
6 Nica AC, et al: Candidate causal regulatory effects by integration of expression QTLs with complex trait genetic associations. PLoS Genet 2010;6:e1000895.
7 Nicolae DL, et al: Trait-associated SNPs are more likely to be eQTLs: annotation to enhance discovery from GWAS. PLoS Genet 2010;6:e1000888.
8 Choi HJ, et al: Patterns of gene and metabolite define the effects of extracellular osmolality on kidney collecting duct. J Proteome Res 2012;11:3816–3828.
9 Cheval L, et al: Atlas of gene expression in the mouse kidney: new features of glomerular parietal cells. Physiol Genomics 2011;43:161–173.
10 Sun Y, et al: Glomerular transcriptome changes associated with lipopolysaccharide-induced proteinuria. Am J Nephrol 2009;29:558–570.
11 Teramoto K, et al: Microarray analysis of glomerular gene expression in murine lupus nephritis. J Pharmacol Sci 2008;106:56–67.
12 Tryggvason SH, et al: A meta-analysis of expression signatures in glomerular disease. Kidney Int 2013;84:591–599.
13 Hodgin JB, et al: A molecular profile of focal segmental glomerulosclerosis from formalin-fixed, paraffin-embedded tissue. Am J Pathol 2010;177:1674–1686.
14 Neusser MA, et al: Human nephrosclerosis triggers a hypoxia-related glomerulopathy. Am J Pathol 2010;176:594–607.
15 Berthier CC, et al: Enhanced expression of Janus kinase-signal transducer and activator of transcription pathway members in human diabetic nephropathy. Diabetes 2009;58:469–477.
16 Sitek B, et al: Novel approaches to analyse glomerular proteins from smallest scale murine and human samples using DIGE saturation labelling. Proteomics 2006;6:4337–4345.
17 Barati MT, et al: Proteomic analysis defines altered cellular redox pathways and advanced glycation end-product metabolism in glomeruli of db/db diabetic mice. Am J Physiol Renal Physiol 2007;293:F1157–F1165.
18 Potthoff SA, et al: The glomerular proteome in a model of chronic kidney disease. Proteomics Clin Appl 2008;2:1127–1139.
19 Lu Y, et al: Bioinformatics analysis of proteomic profiles during the process of anti-Thy1 nephritis. Mol Cell Proteomics 2012;11:M111 008755.

20 Musante L, et al: Characterization of plasma factors that alter the permeability to albumin within isolated glomeruli. Proteomics 2002;2: 197–205.
21 Nakatani S, et al: Targeted proteomics of isolated glomeruli from the kidneys of diabetic rats: sorbin and SH3 domain containing 2 is a novel protein associated with diabetic nephropathy. Exp Diabetes Res doi: 10.1155/2011/979354. Epub 2011.
22 Miyamoto M, et al: In-depth proteomic profiling of the normal human kidney glomerulus using two-dimensional protein prefractionation in combination with liquid chromatography-tandem mass spectrometry. J Proteome Res 2007;6:3680–3690.
23 Yoshida Y, et al: Two-dimensional electrophoretic profiling of normal human kidney glomerulus proteome and construction of an extensible markup language (XML)-based database. Proteomics 2005;5:1083–1096.
24 Waanders LF, et al: Quantitative proteomic analysis of single pancreatic islets. Proc Natl Acad Sci USA 2009;106:18902–18907.
25 Koitabashi K, et al: Micro-sieving: isolation of whole glomeruli from a single renal needle biopsy sample. Nephron Clin Pract 2011; 117:c225–c229.
26 Satoskar AA, et al: Characterization of glomerular diseases using proteomic analysis of laser capture microdissected glomeruli. Mod Pathol 2012;25:709–721.
27 Shapiro JP, et al: A quantitative proteomic workflow for characterization of frozen clinical biopsies: laser capture microdissection coupled with label-free mass spectrometry. J Proteomics 2012;77:433–440.
28 Boerries M, et al: Molecular fingerprinting of the podocyte reveals novel gene and protein regulatory networks. Kidney Int 2013;83: 1052–1064.
29 Kabgani N, et al: Primary cultures of glomerular parietal epithelial cells or podocytes with proven origin. PLoS One 2012;7:e34907.
30 Kang HG, et al: Transcriptome analysis of the response of cultured murine podocytes to puromycin aminonucleoside. Nephron Exp Nephrol 2010;115:e1–e8.
31 Brunskill EW, Georgas K, Rumballe B, Little MH, Potter SS: Defining the molecular character of the developing and adult kidney podocyte. PLoS One 2011;6:e24640.
32 Schordan S, et al: Alterations of the podocyte proteome in response to high glucose concentrations. Proteomics 2009;9:4519–4528.
33 Viney RL, et al: A proteomic investigation of glomerular podocytes from a Denys-Drash syndrome patient with a mutation in the Wilms' tumour suppressor gene WT1. Proteomics 2007;7:804–815.
34 Warsow G, et al: PodNet, a protein-protein interaction network of the podocyte. Kidney Int 2013;84:104–115.
35 Endlich N, et al: Analysis of differential gene expression in stretched podocytes: osteopontin enhances adaptation of podocytes to mechanical stress. FASEB J 2002;16:1850–1852.
36 Jain S, et al: Expression profiles of podocytes exposed to high glucose reveal new insights into early diabetic glomerulopathy. Lab Invest 2011;91:488–498.
37 Saleem MA, et al: The molecular and functional phenotype of glomerular podocytes reveals key features of contractile smooth muscle cells. Am J Physiol Renal Physiol 2008;295:F959–F970.
38 Pavenstadt H, Kriz W, Kretzler M: Cell biology of the glomerular podocyte. Physiol Rev 2003;83:253–307.
39 Ni L, Saleem M, Mathieson PW: Podocyte culture: tricks of the trade. Nephrology (Carlton) 2012;17:525–531.
40 Kopp JB, et al: APOL1 genetic variants in focal segmental glomerulosclerosis and HIV-associated nephropathy. J Am Soc Nephrol 2011;22:2129–2137.
41 Genovese G, et al: A risk allele for focal segmental glomerulosclerosis in African-Americans is located within a region containing APOL1 and MYH9. Kidney Int 2010;78: 698–704.
42 Genovese G, et al: Association of trypanolytic ApoL1 variants with kidney disease in African-Americans. Science 2010;329:841–845.
43 Lipkowitz MS, et al: Apolipoprotein L1 gene variants associate with hypertension-attributed nephropathy and the rate of kidney function decline in African-Americans. Kidney Int 2013;83:114–120.
44 Madhavan SM, et al: APOL1 localization in normal kidney and nondiabetic kidney disease. J Am Soc Nephrol 2011;22:2119–2128.
45 Church DM, et al: Lineage-specific biology revealed by a finished genome assembly of the mouse. PLoS Biol 2009;7:e1000112.
46 Liao BY, Zhang J: Null mutations in human and mouse orthologs frequently result in different phenotypes. Proc Natl Acad Sci USA 2008;105:6987–6992.

47 Ju W, et al: Defining cell-type specificity at the transcriptional level in human disease. Genome Res 2013;23:1862–1873.
48 Uhlen M, et al: Towards a knowledge-based Human Protein Atlas. Nat Biotechnol 2010; 28:1248–1250.
49 Hinkes B, et al: Positional cloning uncovers mutations in PLCE1 responsible for a nephrotic syndrome variant that may be reversible. Nat Genet 2006;38:1397–1405.
50 McMahon AP, et al: GUDMAP: the Genitourinary Developmental Molecular Anatomy Project. J Am Soc Nephrol 2008;19:667–671.
51 Potter SS, Brunskill EW: Building an atlas of gene expression driving kidney development: pushing the limits of resolution. Pediatr Nephrol 2013, Epub ahead of print.
52 Brunskill EW, Potter SS: Changes in the gene expression programs of renal mesangial cells during diabetic nephropathy. BMC Nephrol 2012;13:70.
53 Kretzler M, Cohen CD: Integrative biology of renal disease: toward a holistic understanding of the kidney's function and failure. Semin Nephrol 2010;30:439–442.
54 Barisoni L, et al: Digital pathology evaluation in the multicenter Nephrotic Syndrome Study Network (NEPTUNE). Clin J Am Soc Nephrol 2013;8:1449–1459.
55 Gadegbeku CA, et al: Design of the Nephrotic Syndrome Study Network (NEPTUNE) to evaluate primary glomerular nephropathy by a multidisciplinary approach. Kidney Int 2013;83:749–756.

Matthias Kretzler
University of Michigan, 1570 MSRBII
1150 W. Medical Center Drive
Ann Arbor, MI 48109-0680 (USA)
E-Mail kretzler@med.umich.edu

Author Index

Subject Index